To

Peter

IN MEMORIAM

Contents

Point Counterpoint: A Preface

> It is a commonplace to state that whatever one may come to consider a truly American trait can be shown to have its equally characteristic opposite. . . . a nation's identity is derived from the ways in which history has, as it were, counterpointed certain opposite potentialities; the ways in which it lifts this counterpoint to a unique style of civilization, or lets it disintegrate into mere contradiction.
>
> — Erik H. Erikson

HISTORY and circumstance have endowed Americans with a heritage rich in themes of counterpoint. They began to appear quite early in our history — as soon as paleface confronted redskin. An East-West (or Home-and-Frontier) counterpoint appeared before the North-South variation developed, just as there was a White-Red before there was a White-Black variation. The Old Settler-Immigrant counterpoint turned up about as soon as the second wave of immigrants landed. The Protestant-Catholic and the Gentile-Jewish themes were not long in following.

What America has lacked in the way of overt class conflict she has made up in indigenous tensions of her own peculiar heritage. Tensions of class were undoubtedly present and sometimes irrepressible, but Americans have characteristically thought and acted in terms of regional, religious, racial, or ethnic rather than class conflicts. In fact, Americans have often used them — nowhere oftener than in the South — to avoid or paper over class conflict. Their politics, their compromises, and their institutions have reflected these preoccupations and evasions in much of their history.

More recently, the dissonance in many historic themes of American counterpoint has been resolved by ensuing consonance of some sort. The East-West clash faded in drama and vividness after the West no longer represented a frontier and ceased to constitute a pull away from home. The West then became largely an extension of the East. Old religious tensions have subsided with the secularization of society and the homogenization of sects. Virtually all denominations in America — Protestant, Catholic, or Jewish — now resemble each other, we are told, more than they resemble their Old World mother churches. A half century of the quota system and other restraints on immigration have reduced the visibility of European ethnic minorities made up of first or second-generation immigrants — what used to be called "hyphenate Americans," Irish-Americans, German-Americans, Polish-Americans. Ethnic consciousness still operates, but as Daniel Aaron has pointed out in discussing Jewish-American writers in contemporary literature, the adjective "hyphenate" has dropped out of usage and tends to disappear from dictionaries — just as the ethnic minorities it once designated have been losing their old distinctiveness. The homogenization of minorities has been part of the process of nationalization that has been reducing the distinctiveness of regional and religious groups.

Two huge American minorities, however, have so far largely eluded the great assimilation. They are, ironically, the oldest and (barring the redskins only) the most indigenously American minorities of all. Both of them were established on these shores before the Pilgrims hit Plymouth Rock. They are both the oldest and the latest of the "hyphenates" — the Southern-Americans and the Afro-Americans. "A certain kind of stigma attached to being a white Southerner," says a Negro writer, with tongue only partly in cheek, "just as there is a stigma attached to being a Negro."

These two hyphenate minorities differ markedly, of course, in the origins and character of their alienation and in the doggedness of their dedication to distinctive identity. For the Afro-American, alienation has been from the start involuntary, unavoidable, compulsory. As Daniel Boorstin has put it, the Negro was the "indelible immigrant," unassimilable by definition. In fact, he had completed his immigration before the word "immigrant" came into usage. He has never been accorded the status of immigrant either by the official census or by the historiography of American immigration. To have done so would have been to acknowledge him as a candidate for graduation from that status—to make him potentially assimilable, and that he has been denied. He was, so to speak, a permanent hyphenate. Making a virtue of necessity, he has tended recently to insist on his distinctiveness and sometimes even his alienation and unassimilability — thereby giving a voluntary color to an involuntary plight.

The white Southerner, on the other hand, has been at least in part a hyphenate by choice. Granted the involuntary character of birth and heritage, these are not insuperable obstacles or ineradicable marks. On the one hand the Southern white can, with sufficient determination, minimize, suppress, or wholly deny the distinguishing marks of original identity. And then also, the national threshold of assimilation for Southerners has been distinctly lower in some periods and places than in other periods and places. There have, in fact, been times when Southernism, even some of its rankest manifestations, was scarcely a barrier to acceptance at all, though the threshold would seem to be back up again. The other avenue of assimilation, deliberate divestment of regional identity, has never been the choice of many Southerners and is still the choice of few.

The ironic thing about these two great hyphenate minorities, Southern-Americans and Afro-Americans, confronting each other on their native soil for three and a half centuries, is the degree to which they have shaped each other's destiny, determined each other's isolation, shared and molded a common culture. It is, in fact, impossible to imagine the one without the other and quite futile to try. What the Negro-American would have been without his centuries of Southern experience it is impossible to say, but it is obvious that he would not have been the same. It is not quite so obvious the extent to which the Negro has determined the character and culture of the white Southerners. One reason for that is the common fallacy of identifying everything Southern as white, while forgetting that there is no one more quintessentially Southern than the Southern Negro. In spite of all this, it has been the relations between these two Southern minorities that more than anything else have determined and prolonged the hyphenate distinctiveness of both in American life.

If it is impossible to imagine either of the two old racial components of the South without the other, or the peculiar regional subculture they produced without the combination, it is quite as difficult to conceive of the distinctiveness of American life and national history without the presence of both or without the North-South polarity that presence proclaims and the myths it inspires. The South has been almost as essential to the North and North to the South in the shaping of national character and mythology as the Afro-American to Southern-American and vice versa. North and South have used each other, or various images and stereotypes of each other, for many purposes. They have occasionally used each other in the way Americans have historically used Europe — not only to define their identity and to say what they are *not,* but to escape in fantasy from what they *are.* They have sought in each other images of what they have found wanting at home, for one, the compensatory dream of aristocracy, the airs of grace and decorum left behind, secretly yearned for but never realized. For the other, there were compensatory yearnings for wealth and success and efficiency and metropolitan glamor that were in such short supply at home.

In the main, however, North and South have served each other as inexhaustible objects of invidious comparison in the old game of regional polemics. *Our* faults are as nothing, they have said over and over again, compared with *theirs.* Compare our candor with their hypocrisy, our forthrightness with their evasiveness. Each has served the other as scapegoat for domestic embarrassments and burdens of guilt. Back and forth the dialogue has gone, sometimes at shrill pitch and sometimes in low key, depending on the temper of the times and the moods and needs of the participants. These moods and needs have varied from self-pity to self-righteousness, from guilt and penitence to indignation and aggression. The dialogue, as we shall see, has touched on many historic subjects of controversy. The most prominent subject (often concealing other divisive issues) has been the central and persistent moral problem of American society. Essentially, it has been a white man's dialogue, with the central figure a silent, often indifferent and cynical spectator.

There is need for a history of North-South images and stereotypes, of when and how and why they were developed, the shape they took, the uses that have been made of them and how they have been employed from time to time in regional defense, self-flattery, and polemics. The conventional assumption among scholars as well as laymen is that they were a product of the sectional crisis over slavery and evolved out of the clamor raised by the abolitionists and their opponents. The probability is that they are of remoter origin and have had more varied uses. It is clear that they have outlived the issue that gave them most prominence. All along they have been prolific breeders of regional myth, and their fertility is not yet exhausted.

In addition to intensifying old themes of regional conflict, recent history has imparted new convolutions to the North-South dialogue, new confusions in regional image and stereotype and myth. The migration of millions of Negroes from the South to Northern cities in the last few decades, the greatest internal migration of an ethnic group in American history, has been at the root of these complications. Confronted with an outbreak of problems they associate with the South, Northern political analysts have come to describe them as "the South within the North." Here in the big cities of the North, long considered the very antithesis of the "agrarian" South, has erupted the whole

"Southern" pattern of racial prejudice, violence, and repression in politics, schools, streets, courts, everywhere — hence the "South-within-the-North" formula. This has given birth to a new verb in political science — "to southernize." Thus one well-known analyst speaks of "the effort to southernize the North" and poses a question: "Will the South become more like the rest of the nation or will the rest of the country be southernized?" [1]

A variation of this analysis slightly more charitable to the South takes account of the findings of national commissions on violence and civic disorder that reveal the existence of deep racial prejudices and all their manifestations in the nation at large, North as well as South. Instead of attributing national faults to a regional source, one historian in a rush of magnanimity confesses national involvement in them all. The South is simply America exaggerated, "the essence of the nation" — a distillation of its characteristic faults. The South "crystallizes the defects of the nation." It is seen to be "marvelously useful as a mirror in which the nation can see its blemishes magnified, so that it will hurry to correct them." The mystery of the South's distinctiveness is thus solved: it is different without being distinctive. "It is not a mutation born by some accident into the normal, lovely American family," not a bastard child — only the black sheep. What is true of racism is also found to be true of other unlovely traits such as violence, bigotry, xenophobia, nativism, militarism, chauvinism, and reactionary extremism — they are not Southern peculiarities so much as Southern specialties. The South thus becomes (as metaphors multiply) "a kind of Fort Knox of prejudice — where the nation has always stored the bulk of its bigotry," while circulating the rest somewhat surreptitiously nationwide. It is admitted that concentrating on these as Southern faults tends to gloss over national shortcomings, "but it has the advantage of focusing attention on what is worst." [2]

[1] Samuel Lubell, *The Hidden Crisis in American Politics* (New York, 1970), 86, 88, 141, 186.

[2] Howard Zinn, *The Southern Mystique* (New York, 1964), 217–263.

Another theory, a popular and rather cynical one, attempts to explain the location and intensity of race prejudice and the harshness of race policy entirely in terms of population ratio. Crudely stated, this theory suggests that the larger the percentage of Negroes, the intenser the prejudice and the harsher the policies. This theory tends to give explanation the color of apology and to account for North-South differences or convergencies by reference to determinants beyond the control of either. It serves as a genial rapprochement between sections, an instrument of accommodation for mutual embarrassments. The trouble with this theory is that it does not work. It works neither at home nor abroad as an explanation of race prejudice and policy. It is clear that racial prejudices and policies in other biracial societies of the Americas with much higher percentages of Negro population than the United States, or the South at any period for that matter, were milder in certain ways. It is also clear that the theory does not account for the intensity and distribution of racism in the United States and some other countries. The Middle Western and Northeastern states never had more than one percent Negro population before the Civil War, yet racial prejudices ran to great extremes in those regions and state laws excluding Negroes were common in the Middle West. Similar attitudes and laws existed in Utah and Oregon and other Western territories where Negroes were to be found, if at all, in mere handfuls. The black population of Canada rarely exceeded one percent, yet according to the study by Robin W. Winks their lot in that country bore resemblance to their plight in the republic to the south. It took less than two percent colored of various nationalities to produce a racial crisis in recent British politics. Had those states in the South with a population of fifty percent or more Negroes reacted to the black presence according to the population ratio theory, had they in other words been that many more times as prejudiced, repressive, and severe, the consequences would have been quite unimaginable.

The present work does not attempt to meet the need for a history of the North-South images and stereotypes, nor does it attempt to canvass all the myths of regional identity derived from them. It does treat many of these subjects historically as they have taken various forms over a long span of time — as American history goes. It dips back into colonial history for origins and leaps forward a few times into the twentieth century for a glance at consequences and survivals. The book is divided into roughly equal parts by the Civil War, with five chapters up to that period and five afterward. The major attention throughout is on the nineteenth century.

Sometimes the focus is on the South and sometimes on the North, according to the concentration of disputed issues — and those issues tended to concentrate oftener below the Potomac than above. Wherever the subject of dispute concentrates, however, the disputants tend to range themselves along regional lines — though not always. The most conspicuous disputes along regional lines down through the South's failure to gain its independence clustered around the question of slavery. The first subject studied, "The Southern Ethic in a Puritan World," has roots that go back before slavery and has often been treated without much reference to the institution, though it will be argued here that the influence of slavery was indeed a significant one. "Protestant Slavery in a Catholic World" takes account of a new dimension lately added to the North-South dialogue, the comparative reference extended beyond the traditional interregional scope to include comparisons between the South and other slave societies of the Americas. The broadened perspective has removed many parochial and cultural barriers to understanding, but it will be apparent that it has not removed the subject as a divisive issue from the regional dialogue. The same is true of the old subject of the slave trade, a matter of endless interregional and international debate that is treated in "Southern Slaves in the World of Thomas Malthus." Here again, the comparative approach has greatly enriched understanding, but it has posed new problems that if anything intensify traditional issues over the treatment of slaves in the South.

After slavery came to an end, the focus of the North-South

dialogue shifted to the question of race relations and the status of the Negro. But even before that and while the slavery issue was still under heated debate, the race problem figured powerfully in shaping the antislavery movement. This theme is emphasized in "The Northern Crusade Against Slavery." It is also central to the treatment of Reconstruction in "Seeds of Failure in Radical Race Policy." The three following chapters deal in various ways with race policy and race relations as subjects of intra- as well as intersectional dispute or historiographical controversy. On the latter subject it is hoped that the introduction of comparative dimensions beyond the national limits may prove to be a way of harmonizing conflicting points of view. The same hope is at least part of the motive in treating other controversial themes dealt with in this book.

Counterpoint in music is defined as "a technique of combining two or more melodic lines in such a way that they establish a harmonic relationship while retaining their linear individuality." It seems that harmony is based on dissonance as well as consonance. Applying the technique of counterpoint to such dissonant and prickly lines of thought and controversy as are treated here may rarely be expected to produce a perfect harmonic relationship, for the linear individualities remain strong. To recall Erik Erikson's idea about how a nation derives its identity, quoted at the opening of this Preface, it is hoped at least that we may have helped history a bit in the ways it has "counterpointed certain opposite potentialities; the ways in which it lifts this counterpoint to a unique style of civilization," and that we have not contributed to the ways in which history has permitted counterpoint to "disintegrate into mere contradiction."

I

The Southern Ethic in a Puritan World

MYTHS that support the notion of a distinctive Southern culture tend to be Janus-faced, presenting both an attractive and an unattractive countenance. The side they present depends on which way they are turned and who is manipulating them. The reverse side of Chivalry is Arrogance, and the other side of Paternalism is Racism. The Plantation myth is similarly coupled with the Poor White myth, the myth of Honor with that of Violence. Graciousness, Harmony, and Hospitality also have their less appealing faces. And for Leisure there is its longstanding counterpart, Laziness — together with the synonyms, idleness, indolence, slothfulness, languor, lethargy, and dissipation.

Early and late Southerners and their friends and critics have worried the Leisure-Laziness myth. Alexis de Tocqueville characteristically chose one of the middling terms: "As we advance toward the South," he wrote, "the prejudice which sanctions idleness increases in power." It all began with Captain John Smith's jeremiad against "idleness and sloth" at Jamestown and continued with Robert Beverley's disquisition on industrious beavers and William Byrd's lamentations on the slothful Carolinians of "Lubberland." An impressive literature on the subject has proliferated over the centuries. Leisure, the brighter side of the coin, has been repeatedly praised as an ideal, a redeeming quality of the Southern way that sets it off against the antlike busyness and grubby materialism of the Northern way, another adornment of the Cavalier to shame the Yankee. (Benjamin Franklin voiced the Yankee attitude: "Leisure is the Time for doing something useful.") Laziness, the darker face of the legend, has been deplored, denied, lamented, and endlessly explained. The explanations include climate, geography, slavery, and the staple crop economy, not to mention pellagra and a formidable list of parasites. No matter which face of the myth is favored and which explanation is stressed, however, agreement is fairly general that this is a fundamental aspect of Southern distinctiveness.

As much as they deplored laziness, even exponents of the New South and the Northern way of bustle and enterprise often retained a fondness for the reverse side of the stereotype and clung to leisureliness as an essential mark of regional identity. Although Walter Hines Page scorned many characteristics of the old regime and urged a program of industrialization for the South, he cherished "the inestimable boon of leisure," hoped that "in the march of industrialism these dualities of fellowship and leisure may be retained in the mass of people," and thought that "no man who knows the gentleness and dignity and the leisure of the old Southern life would like to see these qualities blunted by too rude a growth of sheer industrialism." [1]

At the opposite pole from Page on the matter of industrialization for the South, John Crowe Ransom, champion of agrarianism and "the old-time life," also fixed upon leisure as the essence of the Southern ethic. Whatever the shortcomings of the Old South, "the fault of being intemperately addicted to work and to gross material prosperity" was hardly among them, he says. "The South never conceded that the whole duty of man was to increase material production, or that the index to the degree of his culture was the volume of his material prosperity." The arts of the South were "arts of living and not arts of escape." All classes participated in these arts and their participa-

[1] Walter Hines Page, *The Rebuilding of Old Commonwealths* (New York, 1902), 111, 114, 141.

tion created the solid sense of community in the South. "It is my thesis," he writes, "that all were committed to a form of leisure, and that their labor itself was leisurely." [2]

Between Page and Ransom, though much closer in ideology to the former than the latter, W. J. Cash was more ambivalent than either about leisure. Conceding that the South's "ancient leisureliness — the assumption that the first end of life is living itself . . . is surely one of its greatest virtues," he thought that "all the elaborately built-up pattern of leisure and hedonistic *drift;* all the slow, cool, gracious and graceful gesturing of movement . . . was plainly marked out for abandonment as incompatible with success" and the ethic of the new industrial order. This was not to be written off as pure loss. The lot of the poor white had been a "void of pointless leisure," and for that matter, "In every rank men lolled much on their verandas or under their oaks, sat much on fences, dreaming." Leisure easily degenerated into laziness. Both leisure and laziness perpetuated frontier conditions and neither produced cities or a real sense of community. Cash leaned to the Northern view that culture and community are largely developed in towns, "and usually in great towns." [3]

When one gets down to modern economists and the theory of underdeveloped countries and regions like the South, ambivalence about leisure and laziness tends to disappear. The two become almost indistinguishable and equally reprehensible. In the opinion of one Southern economist, leisure becomes "a cover-up for lack of enterprise and even sheer laziness among Southerners." For the mass of them, in fact, "leisure was probably from early days a virtue by necessity rather than by choice. If they were small landowners they lacked sufficient resources to keep them busy more than half of the year. If they were slaves or sharecroppers the near impossibility of advancing themselves by their own efforts bred inefficiency, lassitude, and improvidence."

[2] Twelve Southerners, *I'll Take My Stand: The South and the Agrarian Tradition* (New York, 1930), 12–14.

[3] W. J. Cash, *The Mind of the South* (New York, 1941), 384, 150, 50, 95.

The phenomenon of television aerials over rural slums "may clearly make (through its positive effect on incentives) its own substantial contribution to regional economic progress." But the tradition of leisure has been a major cause of underdevelopment.[4]

I

New light on this ancient dispute comes from a number of sources — new works on moral, intellectual, economic, and slavery history — and a reconsideration of the debate would seem to be in order. One of these new contributions is a work by David Bertelson, whose preference between the Janus faces is clearly announced by the title of his book, *The Lazy South.* The informality of the title might suggest to some that this is another lighthearted and genial essay on regional foibles. Nothing could be further from the author's intentions. He is fully conscious of the "strong volitional and moral connotations" of the word "laziness" and never flinches in his employment of it. This presumes to be the history of a failure, the moral failure of a whole society.

Mr. Bertelson is not one to borrow support casually from conventional assumptions and traditional explanations. Much has been written of geographical determinants of Southern history. "But geography did not create the South," he writes. Pennsylvania, with its waterways and its soil, was as easily adaptable to a tobacco staple culture as Virginia. Nor will he seek an out in economic determinism: rather he asks what determined the economy. He scarcely pauses over the old climatic and ethnic chestnuts. As for the uses of the Peculiar Institution, he is simply not in the market. "Negro servitude did not make the southern colonies different from New England and Pennsylvania. They were different first. That is why slavery became so widespread there. The presence of a small number of slaves in the

[4] William H. Nicholls, *Southern Tradition and Regional Progress* (Chapel Hill, 1960), 34–39.

northern colonies did not change the essential conditions of life in these either. That is why the number remained so small." He has little space for the familiar biological determinants — pellagra, malaria, hookworm, and the rest. They were the accompaniments of poverty and laziness, not their causes. Laziness afflicted the affluent as often as the parasites afflicted the indigent.[5]

What then does account for this distinctive trait of the South? "The difference lay not in the land but in the people," we are told, "and that difference was ultimately due to the different attitudes and assumptions which they brought with them and their descendants perpetuated." The attitudes and assumptions derive ultimately from values, in other words morals, ethics. "To get at what lies behind poverty, slavery, staple crops, and stressing personal enjoyment one must consider historically the meaning of work in the South." It all boils down to an analysis of the Southern Ethic — though that is not a term the author uses.[6]

Virginia existed first in the minds of Englishmen as an answer to the problem of idleness in the mother country. The cure for idleness, it was thought, lay in the allurements of material gain in the New World, which would induce men to work without the necessity of violating their freedom. Spokesmen of Virginia were sometimes vague about personal responsibility to society and a sense of community. Puritans, on the other hand, "thought of themselves as small societies before they established communities." [7] Authoritarians failed in Virginia, and no prophet of a godly community prevailed, no Chesapeake Zion appeared. The same was true of Maryland, the Carolinas, and after abortive experiments, Georgia.

Without coercion and community, Southern colonists established societies based on the exploitation of natural resources. The motive force was individual aggrandizement, we are told, not social purpose or community aims. Southern colonists established plantations, not cities, and cultivated staples, not trade.

[5] David Bertelson, *The Lazy South* (New York, 1967), 244, 104.
[6] *Ibid.*, 244, viii. [7] *Ibid.*, 40–41.

The result was dispersion, fragmentation, and chaotic self-aggrandizement. The meaning of work in the Southern colonies thus came to have no social dimension or content. Work there was, hard work, but it produced idleness with plenty and was intermittent according to seasonal necessities.

"The problem then is," it seems to Mr. Bertelson, "to explain why most men in the southern colonies *seemed* to lead idle lives while in actuality they were often very busy. The answer lies in the fact that they were not busy all of the time." When crops were laid by they tended to go fishing. Max Weber, to whom the author attributes the conception of his study, associated with "the spirit of capitalism" a pattern of industry and diligence that was not based on economic considerations. This pattern is found to be conspicuously absent in the South, along with "the intellectual climate to favor diligence." As a figure of contrast the author pictures "young Benjamin Franklin trudging the streets of Philadelphia in the early morning and his remark that he took care not only to be industrious but to appear industrious." It is conceded that the South was "not without examples of industrious people like Franklin, but . . . most of them were recently arrived foreign Protestants." Your genuine Southerner did not even bother to *appear* industrious.[8]

Yet we are told that the Southerners were deeply troubled with guilt about laziness and took extreme measures to overcome an oppressive sense of purposelessness with sheer "busy-ness." Thus William Byrd, for all his predawn vigils with Hebrew, Greek, and Latin, was desperately driven by the specter of laziness and futility. Even that model of methodical industriousness Thomas Jefferson, in advising his daughter about work habits, "was advocating simply keeping busy rather than purposefulness." Anyway, Jefferson and Madison "expressed very little of the South" in the way, presumably, that Cotton Mather expressed New England and Benjamin Franklin, New England and Pennsylvania. As for George Washington, he proves to have had

[8] *Ibid.*, 75–77.

very rational and Yankeefied notions about work (though the evidence offered is not very persuasive), notions shared by Patrick Henry and Richard Henry Lee. Apart from these unrepresentative deviants, the pattern of Virginia prevailed among Southerners generally. James Winthrop, a Massachusetts Federalist, is quoted as distinguishing sharply between the "idle and dissolute inhabitants of the South" and "the sober and active people of the North." [9]

The trouble with the South, one of the many troubles perceived, was its total inability to conceptualize social unity in terms other than personal relationships, to achieve any real sense of community, or to define industry in social terms. The plantation as community is written off as "imaginary." Southerners, it seems, proved unable to derive social unity "from a community of belief in God and love for one another as it had been for early Puritans." [10] Their efforts to define community in terms of home, personal affections, and local loyalties, the very essence of patriarchal community in the South, are summarily dismissed as futile. Southerners are also said to have been incapable of any real sense of loyalty. Unlike charity, loyalty does not appear to begin at home.[11] "While many Southerners doubtless had great affection for family and friends," he concedes, "and for the localities in which they lived, this did not involve any larger social unity nor any sense of loyalty to the South as a whole — but rather a pervasive particularism." [12]

Personalism and persistent local attachments have been offered by David M. Potter as evidence of the survival of an

[9] *Ibid.*, 166, 161. [10] *Ibid.*, 65.

[11] In a strongly contrasting analysis of the roots of loyalty, David M. Potter writes, "The strength of the whole is not enhanced by destroying the parts, but is made up of the sum of the parts. The only citizens who are capable of strong national loyalty are those who are capable of strong group loyalty, and such people are likely to express this capacity in their devotion to their religion, their community, and their families, as well as in their love of country." "The Historian's Use of Nationalism and Vice Versa," *American Historical Review*, LXVII (July 1962), 932.

[12] Bertelson, *The Lazy South*, 210.

authentic folk culture in the South long after it had disappeared in urban culture elsewhere.[13] While Mr. Bertelson concedes that "at times in the past or in certain limited areas of the South people's lives did attain relatedness and meaning," in the region as a whole disruptive economic forces, pursuit of gain, and chaotic mobility have made impossible "the kind of stability which one usually associates with folk culture." What is "often taken as evidence of a sense of community — courtesy, hospitality, graciousness — is simply a series of devices for minimizing friction only to create the appearance of intimacy or affection." [14] If these particular myths *are* indeed Janus-faced, one face would seem to be effectively veiled from the author's vision.

The emphasis throughout is on the want of any meaningful sense of community in the South. The essence of community is tacitly assumed to be urban, something associated with cities — preferably built on a hill or in a wilderness. "Both the Quakers and to an even greater degree the Puritans in New England founded societies based on communities of consent and common goals. Imbued with a sense of community and social purposefulness, these people were truly able to build cities in the wilderness." [15] It is perhaps inevitable and even appropriate that the myths of an urban society should attach symbolic significance to cities, and it is understandable that these symbols should retain their appeal long after the metropolis has become something less than the ideal embodiment of community, has become in fact the symbol of anticommunity. What is surprising is the total unconsciousness of the mythic quality of these symbols and the faith that they embody.

Harry Levin has commented on the "mythoclastic rigors" of the early Christians. "Myths were pagan," he writes, "and therefore false in the light of true belief — albeit that true belief might today be considered merely another variety of mythopoeic faith. Here is where the game of debunking starts — in the de-

[13] David M. Potter, "The Enigma of the South," *Yale Review,* LI (October 1961), 150–151.
[14] Bertelson, *The Lazy South,* 242–243. [15] *Ibid.,* 244.

nunciation of myth as falsehood from the vantage point of a rival myth." [16]

Mr. Bertelson erects at least one guard against the charge of drawing a geographical line between virtue and vice. He concedes that "Northern society only imperfectly exemplified a sense of community," and that personal aggrandizement, anarchic individualism, and maybe Old Adam himself have been known to break out at times above the Potomac. "The South," he nevertheless maintains, "represents the logical extreme of this tendency. To the degree that America has meant economic opportunity without social obligations or limitations, Southerners are Americans and Americans are Southerners." [17] Some comfort is derived from this remarkable concession, even if it is the comfort the poor white derives in Southern myth from being a repository of the less fortunate traits: at least he belongs.

II

It is, however, just on this score of Southern distinctiveness that a vulnerable spot occurs in the thesis. In a recent analysis of what he calls "the Puritan Ethic," Edmund S. Morgan addresses himself to many of the traditional values that others have denied the South.[18] Yet in Mr. Morgan's analysis these were "the values that all Americans held," even though the claim upon them varied in authenticity, and adherence to them differed in consistency and tenacity. And in spite of the name assigned to the ethic, he disavows any proprietary regional exclusiveness about it and holds that "it prevailed widely among Americans of different times and places," and that "most Americans made ad-

[16] Harry Levin, "Some Meanings of Myth," *Daedalus* (Spring 1959), 225. [17] Bertelson, *The Lazy South,* 245.

[18] Edmund S. Morgan, "The Puritan Ethic and the American Revolution," *William and Mary Quarterly,* 3d. ser., vol. XXIV (January 1967), 3–43. Mr. Morgan will have more to say on the origins of Southern "idleness" in the colonial period that will do more than anything previously published to lift this aspect of the subject out of the speculative stage and give it a foundation of solid research.

herence to the Puritan Ethic an article of faith." [19] Indeed, leaders of North and South found "room for agreement in the shared values of the Puritan Ethic" until the break over slavery. Thomas Jefferson was "devoted to the values of the Puritan Ethic," and Mr. Morgan quotes the identical letters containing Jefferson's advice to his daughter that Mr. Bertelson presents as evidence of the Southern syndrome about idleness and remarks that they "sound as though they were written by Cotton Mather." [20] In a mood of generosity he makes honorary Puritans of numerous Southerners: Richard Henry Lee of Virginia was "a New Englander manqué," Henry Laurens of South Carolina had characteristics that made him "sound like a Puritan," Hugh Williamson of North Carolina was forever "drawing upon another precept of the Puritan Ethic," and the thousands of Southerners who poured into Kentucky and Tennessee in the 1780's "carried the values of the Puritan Ethic with them," from whence they presumably suffused the Cotton Belt.[21]

These references are undoubtedly of generous intent, and there are rules about looking a gift horse in the mouth. Mr. Morgan is careful to say that " 'The Puritan Ethic' is used here simply as an appropriate shorthand phrase" to designate widely held American ideas and attitudes. And perhaps he is right that "it matters little by what name we call them or where they came from." [22] What's in a name? Yet there is *something* about the name that does raise questions. One cannot help wondering at times how some of those Southern mavericks might have reacted to being branded with the Puritan iron. It might have brought out the recalcitrance, or laziness, or orneriness in them — or whatever it was that was Southern and not Puritan. There are in fact, as will later appear, certain limits to the applicability of the Puritan Ethic down South.

By whatever name, however, Mr. Morgan's insights provide needed help in understanding the polemical uses of regional myth and the study of regional character. And he does have a

[19] *Ibid.*, 7, 23, 24, 33, 42. [20] *Ibid.*, 7.
[21] *Ibid.*, 28–29, 38. [22] *Ibid.*, 6.

point that in sectional disputes, each side tended to appeal to the same values. He notes that in the sectional conflict between East and West that followed the Revolution and was accompanied by talk of secession of the lower Mississippi and Ohio Valley, "each side tended to see the other as deficient in the same virtues": "To westerners the eastern-dominated governments seemed to be in the grip of speculators and merchants determined to satisfy their own avarice by sacrificing the interests of the industrious farmers of the West. To easterners, or at least to some easterners, the West seemed to be filled up with shiftless adventurers, as lazy and lawless and unconcerned with the values of the Puritan Ethic as were the native Indians." In this confrontation of East and West and the tendency of each to accuse the other of deficiencies in the same virtues and delinquencies in the same code, Mr. Morgan holds that "the role of the Puritan Ethic in the situation was characteristic." [23]

This sectional encounter of the 1780's recalls another one a century later when the Populists were the spokesmen of Southern (and Western) grievances against the East. The Populist Ethic — perhaps I had better say myth — the Populist Myth had much in common with this Puritan Ethic. According to Populist doctrine, labor was the source of all values, and work to be productive had to have social meaning. Productive labor was the index to the health of a society. Populists swore by producer values. Farmers and laborers were "producers." Merchants were the favored symbol of nonproductive labor. They were "middlemen" who merely moved things around. With them were classified bankers and "monopolists" and "speculators." Their work was not socially useful and served only selfish ends. They exploited the industrious farmers and laborers of the South and West. They deprived men of the just fruits of their labors and removed one of the main motives of industry and frugality. Their gains were ill-gotten and they could be justly deprived of such gains by the state for the welfare of the community, for their profits were not the fruit of virtuous industry and frugality.

[23] *Ibid.*, 20.

Populists were by and large an inner-directed lot, geared to austerity by necessity, suspicious of affluence, and fearful of prosperity. They were the children of a lifelong depression. They looked with baleful eye upon the city as productive of idleness, luxury, extravagance, and avarice. If the city was an appropriate symbol of community for the Puritan Myth, it was just as naturally a symbol of anticommunity for the Populist Myth. Cities were full of nonproducers — merchants, bankers, usurers, monopolists. Merchants and cities were concentrated in the East and particularly in New England. A great deal of Populist animus therefore had a regional target.

The Populists believed they had a direct line of inspiration and continuity from the American Revolution, that Thomas Jefferson was its true spokesman, and that their creed of agrarian radicalism was authentically blessed. They harked back continually to the revolutionary period for fundamental principles and lamented with traditional jeremiads the expiring of republican virtue. It was their historic mission, they believed, to restore virtue and ideals and drive out those who had fouled the temple.[24]

True to tradition, spokesmen of the urban East in the 1890's replied in kind. They pronounced the Populists deficient in virtue and wholly given over to shiftlessness, laziness, and greed. It was the Populists who had defected from the code, not the Eastern capitalists. It was their own deficiencies and lapses from virtue and not exploitation by the East or hostile policies of the government that had brought the farmers to their unhappy plight. Populist resort to government aid and monetary manipulation for relief instead of reliance on frugality and the sweat of their brows were further evidence of shiftlessness and moral delinquency.

Sectional confrontation between the South and other parts of the country in the 1960's hinged more on moral than on eco-

[24] For examples from Populist literature see Norman Pollack (ed.), *The Populist Mind* (Indianapolis and New York, 1967), 51, 66–69, 211–221, 501–519, and *passim*.

nomic issues, but these encounters nevertheless have called forth many familiar recriminations and echoes of the traditional rhetoric of sectional polemics, including those of the 1860's and their deep moral cleavages. Once again, each side tended to view the other as deficient in similar virtues while proclaiming its own undeviating adherence. It is, however, perhaps the first instance in the annals of sectional recrimination (not even excepting the 1860's) in which one region has been seriously denied any allegiance to a common ethic, in which historic violations have been treated not as defections but as the promptings of an alien code. For in this instance the delinquencies were attributed not to some institutional peculiarity that could be eradicated at a cost, nor to some impersonal force of nature or economics for which allowance might be made, but to indigenous, ineradicable attitudes of a whole people "which they brought with them and their descendants perpetuated." This introduces a modern use for an ancient concept resembling the tragic flaw that cursed families in Greek tragedy, something inexorable and fatefully ineluctable.

III

Nothing in this vein of critical evaluation, however, is intended as endorsement of the concept of an undifferentiated national ethic, whether it is called Puritan, Protestant, or by any other name. To be sure, there do exist deep ethical commitments that override barriers of section, class, religion, or race. Otherwise there is no accounting for such limited success as the country has enjoyed in achieving national unity. Mr. Morgan and other historians have served their calling well in bringing these commitments to light. But there still remain sectional differences to be accounted for, and the distinctiveness of the South in this respect is especially unavoidable. The evidence of this distinctiveness, however unsatisfactory the explanations for it may be, is too massive to deny. Where there is so much smoke — whether the superficial stereotypes of the Leisure-Laziness sort, or the bulky literature of lamentation, denial, or celebration that runs back to the seventeenth century, or the analytical monographs of the present day — there must be fire. It remains to find a satisfactory explanation for this aspect of Southern distinctiveness.

Before exploring some explanatory approaches to this problem it would be well to start with an underpinning of agreement, if possible. First there is the question of the foundations of settlement and what the settlers brought with them. Perhaps we could do no better for this purpose than to quote the views of Perry Miller. He pointed out that "however much Virginia and New England differed in ecclesiastical polities, they were both recruited from the same type of Englishmen, pious, hard-working, middle-class, accepting literally and solemnly the tenets of Puritanism — original sin, predestination, and election — who could conceive of the society they were erecting in America only within a religious framework." It is true, he went on to say, that even before Massachusetts was settled, "Virginia had already gone through the cycle of exploration, religious dedication, disillusionment, and then reconciliation to a world in which making a living was the ultimate reality." But even after "the glorious mission of Virginia came down to growing a weed," and even though there were never any Winthrop-type "saints" hanging around Jamestown, the religious underpinnings remained.[25]

In her exhaustive study of Puritanism in the Southern colonies, Babette M. Levy avoids estimates of percentage, but her investigation suggests that a majority of the original settlers were Puritans or Calvinists of some persuasion. Of Virginia she writes that "their presence was felt throughout the colony," and that it was felt extensively in the other Southern colonies as well.[26] Those colonies sustained two further infusions of Puritans, the Huguenots in the seventeenth century and the Scotch-Irish in the

[25] Perry Miller, *Errand into the Wilderness* (Cambridge, Mass., 1956), 108, 138, 139.

[26] Babette M. Levy, "Early Puritanism in the Southern and Island Colonies," in *Proceedings of the American Antiquarian Society*, LXX, Part 1 (1960), 69–348, esp. 86, 119, 308.

eighteenth. They were assimilated in the evolving Southern Ethic, but not without leaving their mark. Then there were the Methodists. Quoting John Wesley to the effect that "the Methodists in every place grow diligent and frugal," Max Weber comments that "the idea of duty in one's calling prowls about in our lives like the ghost of dead religious beliefs." [27] Even after the Enlightenment and the cotton gin those ghosts continued to prowl under the magnolias and, perhaps more wanly, even in the Spanish moss and cabbage palm latitudes. Doubtless they were less at home under the palm than under the pine, but their presence was felt none the less.

Yet it is Weber who points out "the difference between the Puritan North, where, on account of the ascetic compulsion to save, capital in search of investment was always available" and "the condition in the South," where such compulsions were inoperative and such capital was not forthcoming." [28] He goes further to compare religiocultural contrasts in England with those in America. "Through the whole of English society in the time since the seventeenth century," he writes, "goes the conflict between the squirearchy . . . and the Puritan circles of widely varying social influence. . . . Similarly, the early history of the North American Colonies is dominated by the sharp contrast of the adventurers, who wanted to set up plantations with the labour of indentured servants, and live as feudal lords, and the specifically middle-class outlook of the Puritans." [29] Weber was conscious of the paradox that the New England colonies, founded in the interest of religion, became a seedbed of the capitalist spirit, while the Southern colonies, developed in the interest of business, generated a climate uncongenial to that spirit.

This is not the place to enter the "scholarly mêlée" [30] over the

[27] Max Weber, *The Protestant Ethic and the Spirit of Capitalism,* tr. Talcott Parsons (London, 1930), 175, 182.

[28] *Ibid.,* 278n. Sharp exception is taken to this reading of colonial economic history by Gabriel Kolko in his "Max Weber on America: Theory and Evidence," *Theory and History,* I (1961), 243–260.

[29] Weber, *The Protestant Ethic,* 173–174.

[30] For examples see Robert W. Green (ed.), *Protestantism and Capi-*

validity of Max Weber's thesis regarding the influence of the Protestant Ethic in the development of modern capitalism. But here is Weber's delineation of the Puritan Ethic based on his gloss of Richard Baxter, *Saints' Everlasting Rest* (1650) and *Christian Directory* (1673), which he describes as "the most complete compendium of Puritan ethics":

Not leisure and enjoyment, but only activity serves to increase the glory of God, according to the definite manifestations of His will.

Waste of time is thus the first and in principle the deadliest of sins. The span of human life is infinitely short and precious to make sure of one's own election. Loss of time through sociability, idle talk, luxury, even more sleep than is necessary for health, six to at most eight hours, is worthy of absolute moral condemnation. It does not yet hold, with Franklin, that time is money, but the proposition is true in a certain spiritual sense. It is infinitely valuable because every hour lost is lost to labour for the glory of God. Thus inactive contemplation is also valueless, or even directly reprehensible if it is at the expense of one's daily work.[31]

With regard to "the Puritan aversion to sport," unless it "served a rational purpose" or was "necessary for physical efficiency," Weber observes that "impulsive enjoyment of life, which leads away both from work in a calling and from religion, was as such the enemy of rational asceticism, whether in the form of seigneurial sports, or the enjoyment of the dance-hall or the public-house of the common man."[32]

The precise relation of Puritan asceticism to the "spirit of capitalism" is disputed, but in his characterization of the latter, Weber's friend Ernst Troeltsch makes apparent the affinity between the two:

For this spirit displays an untiring activity, a boundlessness of grasp, quite contrary to the natural impulse to enjoyment and ease, and

talism: The Weber Thesis and Its Critics (Boston, 1959), and S. N. Eisenstadt (ed.), *The Protestant Ethic and Modernization: A Comparative View* (New York, 1968).

[31] Weber, *The Protestant Ethic,* 157–158. [32] *Ibid.,* 167–168.

contentment with the mere necessaries of existence; it makes work and gain an end in themselves, and makes men the slaves of work for work's sake; it brings the whole of life and action within the sphere of an absolutely rationalised and systematic calculation, combines all means to its end, uses every minute to the full, employs every kind of force, and in the alliance with scientific technology and the calculus which unites all these things together, gives to life a clear calculability and abstract exactness.[33]

Commenting upon the influence of the Puritan Ethic on "the development of a capitalistic way of life," Weber adds that "this asceticism turned with all its force against one thing: the spontaneous enjoyment of life and all it had to offer." [34]

With all deference to Weber's reputation, this would seem to be an excessively harsh characterization of the Puritan Ethic. He does remind us that "Puritanism included a world of contradictions." [35] Surely the code must have been honored in the breach as well as in the observance, and even Puritans in good standing must have occasional moments of the "spontaneous enjoyment of life." One would hope so, and one is comforted by assurances of modern authorities on the subject that they in fact occasionally did. If not, then it is obvious that "a capitalistic way of life," even with a couple of industrial revolutions thrown in for good measure, came at much too high a cost.

The views of Max Weber, nevertheless, still carry some weight.[36] And if his delineation of the Puritan Ethic bears any

[33] Ernst Troeltsch, *Protestantism and Progress: A Historical Study of the Relation of Protestantism to the Modern World*, tr. W. Montgomery (Boston, 1958), 133–134.

[34] Weber, *The Protestant Ethic*, 166. A mid-nineteenth-century manifestation may be found in James Russell Lowell's *Bigelow Papers:*

> Pleasure doos make us Yankees kind of winch,
> Ez though 't wuz sunthin' paid for by the inch;
> But yit we du contrive to worry thru,
> Ef Dooty tells us thet the thing's to du. . . .

[35] Weber, *The Protestant Ethic*, 169.

[36] For a spirited and persuasive defense of Weber from a recent critic, see Edmund S. Morgan's review of Kurt Samuelsson, *Religion and Economic Action* (New York, 1961), in *William and Mary Quarterly*, 3d. ser., vol. XX (1963), 135–140.

resemblance to the real thing, then it is abundantly clear that no appreciable number of Southerners came up to scratch. There are to be found, of course, authentic instances of Puritan-like behavior in various periods down South. And no doubt a genuine deviant occasionally appeared, some "New Englander manqué," just as a Southerner manqué might, more rarely, turn up in the valleys of Vermont or even on State Street. By and large, however, the great majority of Southerners, including those concerned about their "election," shamelessly and notoriously stole time for sociability and idle talk, and the few who could afford it stole time, and sometimes more than time, for luxury. It is likely that a statistically significant number of them of a given Sunday morning even stole more than their allotted eight hours of sleep. It is extremely unlikely that a sports event — horse race, fox hunt, cock fight, or gander pulling anywhere from the Tidewater to the Delta — was typically preceded by a prayerful debate over whether it "served a rational purpose" or was "necessary for physical efficiency." It was more likely to be contaminated with "the spontaneous expression of undisciplined impulses." Such "inactive contemplation" as went on below the Potomac does not appear to have altered the mainstream of Western civilization, but the temptation to indulge the impulse would rarely have struck the inactive contemplator as morally reprehensible.

A regional propensity for living it up had tangible manifestations of more practical consequence than habits of sociability, sleeping, and sports. Modern economists have often sought to explain why a region of such abundant natural and human resources as the South should have remained economically underdeveloped in a nation of highly developed regional economies — some with poorer natural resources — and suffered the attendant penalties of strikingly lower levels of per capita wealth and income. The Southern scene as of 1930 could be described as "an almost classic picture of an underdeveloped society." [37]

[37] Douglas F. Dowd, "A Comparative Analysis of Economic Develop-

Many explanations are offered and no one of them alone is adequate, but of prime importance in modern theory on the subject is the factor of capital formation. Apart from sheer productivity, the ability to produce more than enough to support the population, the key variable in the rate of capital growth is the willingness to save, or the other side of the coin, the propensity to consume. "Any difference among regions," one economist writes, "in their tendency to consume rather than to save will probably be reflected in their respective rates of capital formation." [38]

Economic and social historians have long remarked on distinctive spending habits, not to say extravagance, of Southerners — the Populist Myth (and reality) to the contrary not withstanding. Most often such habits have been attributed to antebellum planters and their tastes for fast horses, fine furniture, and expensive houses. The flaunting of wealth in an aristocratic society was, according to Weber, a weapon in the struggle for power. These generalizations of historians about extravagance in the South have been based on more or less well-founded impressions and random samples. The only comprehensive statistical study of regional spending habits of which I am aware is one made by several agencies of the federal government covering the years 1935 and 1936 and embracing thousands of families analyzed by population segments and income groups. The limited samples studied indicate that "people in the South did spend, in given income classes, a larger amount for consumption than did residents of other regions"; that in small-city samples low-income groups saved less in the South than in other parts of the country; that at higher income levels in this urban sector "Southerners showed a tendency toward higher levels of consumption and less saving," and that among high farm-income groups in the South "the proportion spent for consumption was

---

ment in the American West and South," *Journal of Economic History*, XVI (December 1956), 558–574.

[38] W. H. Baughn, "Capital Formation and Entrepreneurship in the South," *Southern Economic Journal*, XVI (1949–50), 162.

significantly higher than that for the same farm-income levels elsewhere." [39]

Thus the differences between the Puritan North and the Southern colonies over "the ascetic compulsion to save" that Weber saw operating to influence the availability of capital for investment in the 1600's appeared to be still operative in the 1930's. If the ascetic compulsion is rightly attributed to the Puritan Ethic, then the absence or remission of that compulsion might be another indication of a distinctive Southern Ethic.

IV

The real question then is not whether Southerners fell under the discipline of the Puritan Ethic, but rather — given their heritage and the extent of their exposure — how it was they managed to escape it, in so far as they did. To attribute their deliverance to attitudes they brought with them to America seems rather unhelpful in view of the fact that they brought so many of the same attitudes the New Englanders brought and that these were reinforced by massive transfusions of Puritan blood in later years. Their escape would seem to be more plausibly derived from something that happened to them after they arrived. An acceptable explanation would probably turn out to be plural rather than singular, complex rather than simple, and rather more "environmental" than ideological. To explore and test all the reasonably eligible hypotheses would be the work of elaborate researches and much *active* contemplation. Here it is only possible to suggest a few hypotheses that would appear worth such attention.

Slavery would not come first in any chronological ordering of causal determinants of Southern attitudes toward work, but it hardly seems wise to brush it off entirely into the category of

[39] *Ibid.*, 165–166. Since no allowance was made in these computations for the lower living costs in the South, according to Mr. Baughn, they tend to understate the extent of Southern spending. The validity of these findings are subject to test by further investigation.

consequences. Granting that some distinctive Southern attitudes on work appeared before slavery attained very much importance in the economy, and conceding that these attitudes played a certain part in the spread of the institution, it would be willful blindness to deny the influence that slavery had, once entrenched, in the evolution of the Southern Ethic. The causes of slavery are another subject — a very large one. But once the system became rooted in the land in the seventeenth century, its influence was all-pervasive, and the impact it had on the status of important categories of necessary work and on white and black attitudes toward them was profound and lasting. For not only were these types of work associated firmly with a degraded status, but also fatefully linked with a despised race that continued to perform the same types of work, with no appreciable improvement in racial status, long after slavery disappeared. The testimony of white Southerners themselves, ranging in authority and prestige from Thomas Jefferson to Hinton Rowan Helper, is impressive on the effect that slavery had upon the honor and esteem accorded work in the South. Statistical analysis of the comparative figures on employment by race and by region might be effectively used in testing these impressions.

Regional variations in the nature of work might also deserve attention as determinants of the Southern Ethic. Much of the work required of men in all parts of early America was crude and hard, and little of it anywhere could honestly be characterized as stimulating, creative, or inherently enjoyable. Those who wrote of its joys and rewards probably had a larger share of work that could be so characterized than those who failed to record their impressions. Perhaps it was somewhat easier to surround the labors of shop, countinghouse, and trade with the aura of honor and glamor. That type of work was in short supply in the South.

The agrarian myth of yeoman farmers as "the chosen people of God" certainly did bestow honor and a literary aura of dignity, even glamor, upon a way of making a living, but not necessarily on work itself. No other class of Americans was so persistently and assiduously flattered. The qualities for which the hero of the myth was admired, however, were his independence, his republican virtues, his purifying communion with nature. It was the political and public qualities of the yeomanry that Jefferson had in mind when he wrote that "the small landholders are the most precious part of the state." Work itself was not the essential thing, and certainly not the religion-driven, compulsive, ascetic work of the Puritan Ethic. In fact, neither the hired man nor the slave who did the same work (and probably more of it) shared the dignity and honor conferred by the myth on the yeoman, while the employer or owner of such labor might do little work himself and yet enjoyed the blessings of the myth. In its full flower in the late eighteenth century, the agrarian myth, in fact, stressed escape from the bustle of the world and celebrated pastoral contentment and ease, the blandishments and pleasures of the simple life — the very "impulsive enjoyment of life" held up by Puritan divines as "the deadliest of sins." Never did the agrarian myth regard yeomen as "slaves of work for work's sake." The myth of the happy yeoman was much more congenial to the Southern Ethic than it was to the Puritan Ethic.[40]

As for the black slaves, their very existence did violence to the Puritan Ethic. No one could have a "calling" — in Weber's sense of the term — to be a slave. Since slaves were denied the fruits of their labor, they were deprived of the basic motive for industry and frugality. Circumstance made laziness the virtue and frugality the vice of slaves. Work was a necessary evil to be avoided as ingeniously as possible. Such work as was required of slaves — as well as of those who did work slaves typically performed — was rather difficult to associate with the glory of God or many of the finer aspirations of man. Could such work have been endowed with the mystique of a "calling" or a conviction of divine purpose, it might have rested more lightly on the shoulders of those of whom it was required and on the conscience of those who required it. That seems to have been the way it

[40] Henry Nash Smith, *Virgin Land: The American West as Symbol and Myth* (New York, 1957), 138–150.

worked with onerous nonslave work elsewhere. There appears, however, to have been less stomach for such theological exercises — indeed more need for stomach — in the South than in some other parts. Few sermons to slaves indulged in them. Work therefore lacked the sort of sanctification it might have derived from this source. The attitude of the slave toward work, like many of his attitudes and ways, found secure lodgement in the Southern Ethic.

Turning from the slave to the slaveholder and the plantation system for a look at their influence on the evolution of the Southern Ethic, we move into more troubled waters. Genuine and important differences of opinion about the fundamental nature of the slave economy exist among experts in the field. Members of one school, the most distinguished of whom was Lewis C. Gray, think of the slaveholder as a "planter-capitalist." According to Gray, the plantation system was a "capitalist type of agricultural organization in which a considerable number of unfree laborers were employed under a unified direction and control in the production of a staple crop." [41] The crop was produced for a remote market in response to standard laws of supply and demand by the use of capital sometimes obtained from banks or factors that was invested in land and in slaves. The whole operation was "rational," in the peculiar way economists use that remarkable word — that is, it operated single-mindedly to maximize profits on investment. A contrasting view was that of Ulrich B. Phillips, who believed that the plantation system was not nearly that simple, that it was a complicated "way of life," a social, as well as an economic system. It contained numerous "irrational" elements, goals unrelated or antithetical to the profit motive. It was something of an anachronism, a precapitalist economy existing in a capitalist world.[42]

[41] Lewis C. Gray, *History of Agriculture in the Southern United States to 1860* (2 vols., Washington, 1933), I, 302.

[42] Ulrich B. Phillips, *American Negro Slavery: A Survey of the Supply, Employment and Control of Negro Labor as Determined by the Plantation Regime* (Baton Rouge, 1966), *passim.* The original edition was published in 1918.

It makes a great deal of difference for the implications of the plantation system in accounting for the distinctiveness of the Southern Ethic as to which of these schools is "sound on the goose." For if Gray and his school are right, then there is relatively little to be learned from the political economy of slavery about the distinctiveness of the South from the rest of the uniformly capitalist America. If Phillips is on the right track, however, there might be a great deal to learn. It may be a long time before a consensus on the slave economy is reached among scholars, and certainly no attempt is made to settle the dispute here.

It is instructive to note, however, a convergence between the old-school views of Phillips and a new-school view developed by the investigations of Eugene D. Genovese.[43] In his findings, the plantation economy was shot through with "irrationality" (in the marketplace sense) and given to wild deviations from the capitalistic norm. "The planters were not mere capitalists," he writes; "they were precapitalist, quasi-aristocratic landowners who had to adjust their economy and ways of thinking to a capitalist world market. Their society, in its spirit and fundamental direction, represented the antithesis of capitalism, however many compromises it had to make." He elaborates on the antithesis as follows:

> The planters commanded Southern politics and set the tone of social life. Theirs was an aristocratic, antibourgeois spirit with values and mores emphasizing family and status, a strong code of honor, and aspirations to luxury, ease, and accomplishment. In the planters' community, paternalism provided the standard of human relationships, and politics and statecraft were the duties and responsibilities of gentlemen. The gentlemen lived for politics, not, like the bourgeois politician, off politics.
>
> The planter typically recoiled at the notions that profit should be the goal of life; that the approach to production and exchange should be internally rational and uncomplicated by social values;

[43] See for example Genovese's foreword to the 1966 edition of Phillips, *American Negro Slavery* cited in note 42.

that thrift and hard work should be the great virtues; and that the test of the wholesomeness of a community should be the vigor with which its citizens expand the economy. The planter was no less acquisitive than the bourgeois, but an acquisitive spirit is compatible with values antithetical to capitalism. The aristocratic spirit of the planters absorbed acquisitiveness and directed it into channels that were socially desirable to a slave society: the accumulation of slaves and land and the achievement of military and political honors. Whereas in the North people followed the lure of business and money for their own sake, in the South specific forms of property carried the badges of honor, prestige, and power.[44]

He goes on to observe that "at their best, Southern ideals constituted a rejection of the crass, vulgar, inhumane elements of capitalist society. The slaveholders simply could not accept the idea that the cash nexus offered a permissible basis for human relations." The planters reinforced their paternalism toward their slaves by a semipaternalism toward their neighbors and "grew into the closest thing to feudal lords imaginable in a nineteenth-century bourgeois republic." [45]

Mr. Genovese has greatly complicated and enhanced the fascination of the game of ethic identification. One shrinks at the prospect of encouraging studies of "feudal" and "aristocratic" themes in Southern history, and even more at the anticipated glances from colleagues in European history. It would be well to emphasize the prefix *quasi* that Mr. Genovese judiciously attaches to these terms. With these precautions (and misgivings) and with no disposition to prejudge the outcome of the revived and flourishing controversy over the political economy of American slavery, one might at least say that this interpretation of the slave South contains a number of suggestive hypotheses relating to the distinctiveness of the Southern Ethic.

On the themes of precapitalist economies, the work of Max Weber again becomes relevant. He strongly emphasized the "ir-

[44] Eugene D. Genovese, *The Political Economy of Slavery: Studies in the Economy and Society of the Slave South* (New York, 1965), 23, 28.
[45] *Ibid.*, 30.

rational" characteristics of slave economies, particularly in master-slave relationships, consumer behavior, and politics. At the risk of reviving the Waverly-novel approach, the following characterization of the aristocratic ethic by Weber is offered as summarized by his biographer:

> In feudal ideology the most important relations in life are pervaded by personalized ties, in contrast to all factual and impersonal relationships. . . . From this standpoint luxury is not a superfluous frill but a means of self-assertion and a weapon in the struggle for power. This antiutilitarian attitude toward consumption was of a piece with the equally antiutilitarian orientation toward one's life. Aristocratic strata specifically rejected any idea of a 'mission in life,' any suggestion that a man should have a purpose or seek to realize an ideal; the value of aristocratic existence was self-contained. . . . Aristocrats deliberately cultivated a nonchalance that stemmed from the conventions of chivalry, pride of status, and a sense of honor.[46]

Social role-playing is a broad mark for satire, but all societies engage in it consciously or unconsciously. The sharpest ridicule is reserved for the unfortunate society that is caught at the height of collective posturing, brought low with humiliating exposure of its pretenses, and forced to acknowledge them — or live with them. The Old South has had its share of exposure along this particular line, and more at this juncture has rather low priority among the pressing tasks of historiography. More needful are analytical appraisals of the social content of the patriarchal, paternalistic, and aristocratic values and the remarkable qualities of leadership they developed. A comparative approach could be helpful, but the traditional moralistic comparison with the contemporary society to the north might profitably be exchanged for comparisons with English and Latin slave societies to the south that shared more of the South's traditions, institutions, and values.

An abhorrence of slavery and an identification with abolitionists on the part of both liberal and radical historians have

[46] Reinhard Bendix, *Max Weber, An Intellectual Portrait* (New York, 1962), 364–365.

skewed and clouded their interpretation of the Old South. In a critique of Marxian and liberal historiography of slavery, Eugene Genovese holds that the main problem "arises from the duality inherent in a class approach to morality" and contends that both liberals and Marxists have made the mistake of judging the ruling class of Southern planters "by the standards of bourgeois society or by the standards of a projected socialist society." With references to rulers of the Old Regime, he continues:

These men were class conscious, socially responsible, and personally honorable; they selflessly fulfilled their duties and did what their class and society required of them. It is rather hard to assert that class responsibility is the highest test of morality and then to condemn as immoral those who behave responsibly toward their class instead of someone else's. There is no reason, unless we count as reason the indignation flowing from a passionate hatred for oppression, to withhold from such people full respect and even admiration; nor is there any reason to permit such respect and admiration to prevent their being treated harshly if the liberation of oppressed people demands it. The issue transcends considerations of abstract justice or a desire to be fair to one's enemies; it involves political judgment. If we blind ourselves to everything noble, virtuous, honorable, decent, and selfless in a ruling class, how do we account for its hegemony? . . . Such hegemony could never be maintained without some leaders whose individual qualities are intrinsically admirable.[47]

Also needed are discerning assessments of the skill, conviction, and zest which other players brought to the colorful variety of roles assigned them by the Old Regime. That of the slave has been subjected to appreciative analysis of late, with efforts to illuminate the personality and values of his descendants. The roles of his master and mistress, Lord and Lady Bountiful, as well as those of the lesser gentry, the squirearchy, the yeomanry,

[47] Eugene D. Genovese, "Marxian Interpretations of the Slave South," in Barton J. Bernstein (ed.), *Toward a New Past: Dissenting Essays in American History* (New York, 1968), 114.

and the poor whites deserve comparable study. It is possible that a majority of the players identified completely with their roles as "real life" — as completely perhaps as the saints, prophets, and come-outers in other quarters found identity in their roles. At least the cast for the Old South drama, slaves included, often acted as if they did, and sometimes they put on rather magnificent performances. Study of the institutional setting in which they performed — the patriarchal tradition, the caste system, the martial spirit, the racial etiquette, the familial charisma — all deserve attention from the historian of the Southern Ethic.

V

After the curtain fell on the Old South in 1865, the same cast of characters had to be taught strange roles and learn new lines. For a people who had been schooled so long in the traditional roles, and especially for those who played them for "real life," this was not an easy assignment. The social dislocations and traumas of Reconstruction and the period that followed can be seen as drastic experiments from two sides: the conservative side trying to preserve the Southern Ethic, and the radical side trying to destroy it. As we shall see, neither of these experiments was destined wholly to succeed.

To convert the ex-slaves miraculously into "slaves of work for work's sake" in the classic model of Weber's "spirit of capitalism" was clearly beyond the reasonable expectations of either side. Given the freedman's age-long indoctrination in a work ethic appropriate to his enslavement, and given the necessity of his now doing the same kinds of work without the old compulsions, the question was whether he could be induced to do any work beyond what would provide him a bare subsistence, if that. Planter conservatives were convinced that the only answer was force in some guise, and in lieu of bondage they put forward schemes of apprenticeship and vagrancy laws embodied in "Black Codes." In effect they offered the old allegiance of paternalistic responsibility for abject dependents while tacitly retaining the sanction of the whip.

Ruling these schemes illegal and a travesty on freedom, Northern agencies, including the Freedmen's Bureau, missionaries, and private speculators, confidently maintained that "normal" economic incentives were the solution: wages, profits, and assurance of the fruits of one's labor. Northern agencies and employers launched their program with a campaign of indoctrination laced with the rhetoric of diligence, frugality, and the sanctity of contracts — staples of the Puritan Ethic. Needless to say, the Puritan Ethic had acquired its own Janus face over the centuries, and it was the face of Yankee acquisitiveness that was presented to the South. Along with this went the promise of free land held out to the freedmen by Congress and the Freedmen's Bureau. But that promise ran afoul two conflicting principles of their own: the sanctity of property and the doctrine of equal rewards for equal labor. The first blocked the confiscation of planters' property to provide the free land, and the second inhibited the granting of special privilege unearned by "honest" labor. The North reneged on the promise of free land, and the freedmen sulked. The "normal" incentives were not operating. Northern business joined Southern planters in demanding that the Freedmen's Bureau get the Negro back to the cotton fields. When General O. O. Howard, head of the Bureau, received "authentic complaints of idleness for which no remedy seemed to exist," he ordered enforcement and extension of state vagrancy laws. His assistants withheld relief, compelled freedmen to accept labor contracts, and enforced their harsh terms with penalties of forfeited wages and withheld rations. Gradually the North withdrew and left the freedmen to make what terms they could with planters.[48]

[48] Willie Lee Rose, *Rehearsal for Reconstruction: The Port Royal Experiment* (New York, 1964), 212–216, 308–310; George R. Bentley, *A History of the Freedmen's Bureau* (Philadelphia, 1955), 79–86; for a revisionary estimate of the Bureau see William McFeely, *Yankee Step-*

In the meantime the campaign to convert Southern white men went forward briskly and at first hopefully. "It is intended," declared Thaddeus Stevens, "to revolutionize their principles and feelings," to "work a radical reorganization in Southern institutions, habits, and manners." [49] The difficulties of the undertaking were acknowledged. "The conversion of the Southern whites to the ways and ideas of what is called the industrial stage in social progress," wrote Edwin L. Godkin, was a "formidable task." He believed that "the South, in the structure of its society, in its manners and social traditions, differs nearly as much from the North as Ireland does, or Hungary or Turkey." [50] But the revolutionizing process would go forward, promised Horatio Seymour, "until their ideas of business, industry, money making, spindles and looms were in accord with those of Massachusetts." [51] The Southern whites, like the Negroes, were subjected to indoctrination in diligence, austerity, frugality, and the gospel of work. They were advised to put behind them the "irrationality" of the Old Order, outmoded notions of honor, chivalry, paternalism, pride of status, and noblesse oblige, together with all associated habits of indolence, extravagance, idle sports, and postures of leisure and enjoyment. In the name of "rationality" they were adjured to get in there like proper Americans and maximize profits.

The verbal response from the South, after a refractory interlude and apart from a continued undertone of muttered dissent, must have been gratifying. The antebellum business community, long inhibited by ties to the Old Order, burst into effusions of assent and hosannas of delivery. "We have sowed towns and cities in the place of theories and put business above politics," announced Henry W. Grady to the cheering members of the

father: General O. O. Howard and the Freedmen's Bureau (New Haven, 1968).

[49] Quoted in Howard K. Beale, *The Critical Year: A Study of Andrew Johnson and Reconstruction* (New York, 1930), 149.

[50] *The Nation,* XXXI (1880), 126.

[51] New York *Herald,* October 31, 1866.

New England Society of New York. "We have fallen in love with work." [52] And Richard H. Edmunds of the *Manufacturers' Record,* invoking the spirit of Franklin, announced that "the South has learned that 'time is money.' " [53] A Richmond editor rejoiced that "the almighty dollar is fast becoming a power here, and he who commands the most money holds the strongest hand. We no longer condemn the filthy lucre." [54] The range of New South rhetoric left unsounded no maxim of the self-made man, no crassness of the booster, no vulgarity of the shopkeeper, no philistinism of the profit maximizer. For egregious accommodation and willing compliance the capitulation was to all appearance complete.

But appearances were deceptive, and the North was in some measure taken in by them. The New South orators and businessmen-politicians who took over the old planter states gained consent to speak for the region only on their promise that Southerners could have their cake and eat it too. That was one of the meanings of the Compromise of 1877. The South could retain its old ways, the semblance or outer shell at any rate, and at the same time have an "industrial revolution" — of a sort, and at a price. White supremacy was assured anyway, but home rule and states rights meant further that the South was left to make its own arrangements regarding the plantation system, the accommodation of the freedmen and his work ethic, and his status as well as that of white labor in the new economy and polity. The price of the "industrial revolution" and the reason for the distorted and truncated shape it took was an agreement that the North furnish the bulk of investment capital and entrepreneurship, while the New South managers smoothed the way by a "cooperative spirit" that assured generous tax exemptions, franchises, land grants, and most important of all, an abundant supply of cheap, docile, and unorganized labor.

[52] Quoted in Raymond B. Nixon, *Henry W. Grady: Spokesman of the New South* (New York, 1942), 345.
[53] Baltimore *Manufacturers' Record,* XIV (November 3, 1888), 11.
[54] Richmond *Whig and Advertiser,* April 4, 1876.

The South got its railroads and mines, its Birmingham heavy industry and its Piedmont factories. That is, they were located in the South. But they were largely owned elsewhere and they were operated in the interests of their owners. They were of one general type: low-wage, low value-creating industries that processed roughly the region's agricultural and mineral products but left the more profitable functions of finishing, transporting, distributing, and financing to the imperial Northeast. The South remained essentially a raw-material economy organized and run as a colonial dependency. On the agricultural side a sort of plantation system survived along with the one-crop staple culture, but the absentee owners dropped the wage contracts, substituted debt slavery for chattel slavery, and organized the labor force, now more white than black, as sharecroppers or tenants bound by an iron crop lien into virtual peonage. The cropper-tenants rarely handled any money, subsisted on fatback, cornbread, and molasses, and constituted a mass market for Southern manufacturers of chewing tobacco, moonshine whiskey, and very little else. They were in fact a nonmarket.

The New South neither preserved the Southern Ethic intact nor abandoned the allegiance entirely. It substituted a compromise that retained a semblance at the expense of the essence. The substitute was more defiantly proclaimed, articulately defended, and punctiliously observed than the genuine article had been. The cult of the Lost Cause covered the compromise with a mantle of romantic dignity and heroism. Surviving heroes in gray became Tennysonian knights. The Plantation Legend took on a splendor that shamed antebellum efforts. There was a place in the new cast — with important exceptions to be noted — for all players of the old roles as well as for newcomers with only imaginary identification with them, and there was a gratuitous upgrading of status all around. Great energy went into the performance. Spectators were impressed with the graciousness and hospitality, the leisurely elegance and quaint courtliness at upper levels, and with a pervasive kindliness, a familial warmth, and a deferential courtesy that prevailed generally.

There were even those who affected the patriarchal and paternalistic roles. Right here, however, came the sad falling off, the point at which the sacrificed essence — aristocratic obligations of noblesse oblige, responsibilities of leadership, solicitude for dependents and subordinates, and an antibourgeois disdain for the main chance and the fast buck — gave way to the bourgeois surrogate. For the children of the new patriarchs and the dependents of the new paternalists were the mass of forlorn croppers and tenants, black and white, in their miserable rural slums, undefended victims of ruinous rents and interest rates, mined-out soil, outmoded techniques, debt slavery, and peonage. These and the first-generation industrial proletariat, white and black, in their urban or company-town slums, were victims of the lowest wages, longest hours, and deadliest working conditions in the country.

The oppressed were not so docile as to submit without show of resistance. As the nineties came on, Negroes and white labor under the leadership of desperate farmers combined in the Populist Revolt to mount the most serious indigenous political rebellion an established order ever faced in the South. In their panic over the rebellion, the New South leaders relinquished their last claim to responsibility derived from the Old Order. Abandoning their commitment to moderation in racial policy, they turned politics over to racists. In the name of white solidarity and one-party loyalty, they disfranchised the Negro and many lower-class whites and unleashed the fanatics and lynchers. In their banks and businesses, in their clubs and social life, as well as in their inner political councils, they moved over to make room for increasing delegations of Snopeses. In place of the old sense of community based on an ordered if flexible hierarchy, they substituted a mystique of kinship, or clanship, that extended the familial ambit spuriously to all whites — and to such Negroes as had mastered and sedulously practiced the Sambo role to perfection.

There have been lamentations for the passing of the Southern Ethic all down the years and periodic jeremiads for its demise.

Referring to the Puritan Ethic, Edmund Morgan writes that "it has continued to be in the process of expiring," that it has "always been known by its epitaphs," and that "perhaps it is not quite dead yet." [55] The future of either of these two historic relics in an affluent, consumer-oriented society of the short work week, early retirement, and guaranteed income looks rather dubious, though in a giant supermarket world the ethic of grasshopper would seem to have somewhat more relevance than that of the ant. And Puritanical condemnations of leisure seem a bit quaint. Whether the Southern Ethic is dead or not, the lamentations and jeremiads show no sign of languishing. In fact, the works of William Faulkner constitute the most impressive contribution to that branch of literature yet. If epitaphs are indeed a sign of life in this paradoxical field of ethical history, then the lamentations here, too, may be premature.

[55] Morgan, "The Puritan Ethic" (cited in note 18), 42–43.

2

Protestant Slavery in a Catholic World

THE abolitionist critique of Southern society was elaborate and many-sided, but it was for the most part culture-bound. It was criticism directed by one segment of a society at what was assumed to be another segment of the same society. Both segments, that of the critics as well as that of the criticized, were presumed by abolitionists to share the same framework of values: Protestant, democratic, Anglo-Saxon — morals according to Calvin (with modifications) and politics according to Locke (with more of the same). And so they did historically and, up to a point, contemporaneously. The society of the South was slow in coming to full consciousness of its slave-based ethos and its departure from historical values. George Fitzhugh and his friends only burst into print in the last years of the Old Regime. Enough of the old values remained to give the abolitionist critique a stinging relevance. That was why it was so bitterly resented: it was so often on target.

The burden of abolitionist attack was that the South was hypocritical. It did not practice what it supposedly professed and preached. It subscribed, like all Americans, to the doctrine that "all men are created equal," and then enslaved men and scorned equality; it swore by democracy and exercised absolute power; it sanctioned marriage as holy and prohibited marriage in practice; it avowed the sacredness of the family and refused to respect family ties; it professed respect for property and denied the right to hold property; most loudly of all it preached Anglo-Saxon purity of race, but quietly practiced miscegenation. Abolitionists summed up their indictment by pronouncing slavery a "sin" and renouncing all compromise with the institution in all its variations.

It would not have occurred to the American abolitionists, with their Puritan background and their attitude toward sin, to have grounded their indictment of slavery in the South upon a comparison with slavery in Latin America. To have done so would have been to admit a relativity about sin and slavery abhorrent to their creed. It would have obliged militant Protestants to acknowledge some measure of respect for Catholicism, for Thomist doctrine, for casuitical justice, for a structural corporative society, for tolerance of sin in the human community, and for a host of late-medieval values and notions.[1] It would have also obliged uncompromising democrats and antimonarchists to hold up as examples to fellow Americans the works of Latin aristocrats and some of the most absolute monarchs in operation.

This was unthinkable. And yet within a century after the triumph of the Puritan abolitionists and agreement on the validity of their indictment of American slavery, even by many Southerners, the case has been reopened along the Catholic-Latin front. The prosecution this time was in the hands of secular scholars, and the brunt of the attack was not, of course, on slavery as such but on the long term effects of the American *type* of slavery upon American race relations and the Negro American personality and values. The Anglo-American system of slavery was pictured as almost completely unrestrained by the intervention of church or state and unchecked by tradition. Such laws as governed the rights and conditions of slaves were the work of assemblies consisting of slaveholders themselves. They defined the slaves as chattels, as "things," denied them a moral personal-

[1] Richard M. Morse, "The Strange Career of 'Latin-American Studies,'" *Annals of the American Academy of Political and Social Science*, CCCLVI (November 1964), 106–112.

ity, diminished or virtually ignored their human status, and left them at the mercy of their masters without protection of their most elementary rights. Slavery was assumed to be, in the marketplace sense, a "rational" institution. In the human sense it was characterized as so oppressive, so hostile to freedom, that it left its victims crippled in mind, spirit, and personality and wholly unprepared for the status of freedom and citizenship when it eventually came.

By contrast, slavery in Latin America was pictured as subject to many humanizing restraints and traditions. The continuity of this institution in Spain and Portugal preserved laws of classical Rome and the wisdom of the Middle Ages in slave codes for the New World. Powerful monarchs with vigilant viceroys and royal bureaucracy intervened between master and slave. And so did the Church of Rome, and in combination with the Crown, so did the Holy Inquisition. Slave and master were equal in the sight of God. Not only was the Negro slave endowed with a moral personality, he was accorded the blessings of the seven sacraments and embraced by the powerful church. Elaborate laws protected his life and limb, his marriage and family, and opened avenues to freedom, opportunity, and fulfillment. The line between freedom and slavery became blurred, and so did the line between one race and another. In an atmosphere friendly to freedom, slavery became a school for freedmen and a preparation for racial equality and full assimilation.

The Puritan-abolitionist case against the South was a limited indictment for violations of what Northern abolitionists assumed to be a common code and disrespect for mutual ideals and traditions. (Southerners were only beginning to question the foundations of the common order.) On the other hand, the Catholic, Latin American attack on Protestant Anglo-American slavery is an indictment of the civilization itself, North as well as South, together with its philosophical and political underpinnings. Included under the indictment are those proud colonial assemblies, the first representative governments in the New World, their stiff-necked defiance of monarchy and imperial control, and their responsiveness to the will of the electorate. Included also is the separation of church and state and a clergy that defied the Mother Church, abandoned the immemorial wisdom of Rome, and submitted to the will of wordly parishioners. Called before the bar for judgment are not only Calvin and Locke, but also Adam Smith and his laissez-faire economy, as well as Thomas Jefferson and his secularized state.

The thoroughgoing nature of this oblique contemporary attack on American institutions and credos would seem to present formidable enough obstacles to acceptance of the Catholic-Latin picture as a mirror in which to discover historic Anglo-American imperfections. But there were also additional obstacles rooted in the Black Legend of Spanish America that dates from centuries of English rivalry with Spain and Protestant propaganda against the Pope. The Black Legend pictured a Spain of unmatched cruelty and perfidy, dwelt on ruthless extermination of natives and on blood-stained treasure, on the persecution of Protestants, the horrors of the Inquisition, and the refined tortures of Torquemada. John Foxe's *Book of Martyrs* (1563) was standard Puritan reading and a rich repository of the legend. But long after the martyrs were forgotten and fires of the Reformation cooled, North Americans tended to look upon Latin Americans in a patronizing and superior manner as mixed breeds addicted to revolution and dictators and suffering from "underdevelopment."

In spite of this, Americans in recent years exhibited a surprising willingness, at times an eagerness, to see their faults — historic as well as contemporary — mirrored faithfully in reverse in a Latin American picture of racial felicity and harmony. The disposition became especially pronounced in the 1950's and 1960's and might well have been influenced by the mood of guilt about American race relations that was intensified in those decades by agitation. But it was not only that, for there were obvious differences between the two cultures in race relations and some of them assuredly reflected discredit on the Anglo-American record and the contemporary social scene. For in place of the racial harmony, tolerance, color blindness, ethnic mobility, and felicitous assimilation attributed to happy islands in the sun and republics of the southern continent, American reformers saw all around them evidence of racial disharmony, intolerance, discrimination, segregation, and violence. Surely, these Catholic-Latin societies were proof that race relations could be different, that more civilized policies could prevail, that somewhere in the past Anglo-America took the wrong turn, and that perhaps race conflict and prejudice were unhappy national peculiarities.

These were "lessons learned," not necessarily "lessons taught." Serious historians of Latin American slavery and race relations could not fairly, nor always, be held responsible for inferences drawn by social moralists. Sociologists were especially attracted to this line of thought and invidious comparison. One of them wrote that West Indian blacks were "not ashamed of their color. They do not undertake the impossible in trying to change their features with hair-straightening and skin-bleaching processes, as is so common among American Negroes," and that "Negroes in Brazil have never considered themselves as an inferior minority group. There are no restrictions whatsoever of a caste nature and no limitations along racial lines." [2] Others projected their sociological insights into the past to effect striking historical revisions. "American slavery," wrote one sociologist in an official government report, "was profoundly different from, and in its lasting effects on individuals and their children, indescribably worse than, any recorded servitude, ancient or modern." [3] No American institution has ever suffered such a sweepingly derogatory characterization as this at the hands of an historian, and even slavery at the height of abolitionist agitation was spared such extremes.

Such a reading of the slavery experience as oppressive and

[2] Maurice R. Davie, *Negroes in American Society* (New York, 1949), 401, 460.

[3] [Daniel Patrick Moynihan] U.S. Department of Labor, *The Negro Family: The Case for Federal Action* (Washington, D.C., 1965), 15.

cruel beyond comparison with any other in the history of that ancient institution carried with it certain unhappy implications, not only about the white architects and perpetrators of the system, but also about its victims and their descendants, the Negro Americans. For the "Children of Oppression" could only be assumed to be permanently scarred by such an unprecedented experience of their forebears and to bear evidence of these scars in deficiencies of personality, responsibility, sexuality, mating and childbearing, and family stability. The correlation of historic trauma with sociological data had the charitable impulse of "explaining" and, in a measure, justifying traits often attributed to inherent racial characteristics and at the same time rationalizing modern paternalistic programs of amelioration. Some Negro Americans found the cost too high and resisted this opprobrious characterization of their heritage and their culture as being too debasing in its implications. But others accepted it as a means to worthy ends, if not actually valid historically. White acceptance, less encumbered by considerations of racial pride and self-esteem and able to rise in the liberal tradition above old stereotypes about Catholics and Latin Americans, was more uninhibited and uncritical. Even some Southern whites spoke as if they found nothing essentially incompatible between the new critique of American slavery and the estimate they entertained regarding their own forebears and heritage.

Refinements of the comparative estimates of slavery in the New World extend beyond the broad distinction between Catholic and Protestant or Latin and Anglo-American variants. Several writers have suggested multiple gradations of benevolence according to the reputation of the colonizing power. They differ somewhat in detail, but there is general agreement on three categories that place Portugal and Spain as the most humane, France in the second category, and the Dutch, Danish, and English at the bottom of the list as the most inhumane and oppressive.[4] It is conceded by some that important differences exist

[4] Frank Tannenbaum, *Slave and Citizen: The Negro in the Americas* (New York, 1946), 65n; Stanley M. Elkins, *Slavery: A Problem in*

between Latin American colonies, even between individual colonies of one nation, Latin or not, in the condition of slaves and race relations, and within one colony between one period and another. The more that is written in this relatively new field of comparative slavery, the more significant those variables of time and place appear. Much scholarly energy and genius will be needed to refine points of similarity and contrast and delineate significant differences arising out of time and place.[5]

In the meantime, several considerations suggest Brazil as a suitable point of departure for exploring some of the implications of the Catholic-Latin American critique of Protestant-Anglo-American slavery. Brazil enjoys perhaps the highest reputation in the Americas for its record in slavery and race relations. It has the largest Afro-American population, second only to that of the United States, and the longest experience with slavery, barring none. Furthermore, there exists more scholarly literature on these subjects in Brazil than in any other country except the United States.[6] Of special significance in their litera-

American Institutional and Intellectual Life (Chicago, 1959), 77n; David B. Davis, *The Problem of Slavery in Western Culture* (Ithaca, N.Y., 1966), 57–61.

[5] Eugene D. Genovese, "The Treatment of Slaves in Different Countries: Problems in the Applications of the Comparative Method," in Laura Foner and Eugene D. Genovese (eds.), *Slavery in the New World: A Reader in Comparative History* (Englewood Cliffs, N.J., 1969), 202–210.

[6] The São Paulo School, headed by Florestan Fernandes, whose seminal book *The Negro in Brazilian Society* (New York, 1969) has recently been translated and published in the United States, challenges many of Freyre's ideas. Fernandes and his colleagues Octavis Ianni and Fernando Henrique Cardoso have not yet, however, had the influence in North America that Freyre has long enjoyed. On this influence see Lewis Hanke, "Gilberto Freyre: Brazilian Social Historian," *Quarterly Journal of Latin American Relations,* I (July 1939), 24–44; and Stanley J. Stein, "Freyre's Brazil Revisited: A Review of *New World in the Tropics: The Culture of Modern Brazil,*" *Hispanic American Historical Review,* XLI (1961), 111–113. In addition has been the revisionary influence of C. R. Boxer, especially in *The Golden Age of Brazil, 1695–1750: Growing Pains of a Colonial Society* (Berkeley, California, 1964) and *Race Relations in the Portuguese Colonial Empire, 1415–1825* (Oxford, England, 1963). Freyre himself has shifted gradually to the right since 1950.

ture, because of its influence in the United States and the reputation of the author, is the work of Gilberto Freyre. His big book, *The Masters and the Slaves,* and its sequel, *The Mansions and the Shanties,*[7] have done more than anything else to inspire the revisionary estimate of slavery and race relations in the United States in the light of the Latin American experience. "American historians," writes an English historian, "have a similar but more difficult task to undertake; but they may take heart from the fact that under the influence of men like Freyre, the Brazilians have been so much more successful than themselves in blotting out the idea of race in the minds of the people."[8] The recognition of his influence and the extent of his acclaim in this country were signaled by a message of congratulations from the President of the United States upon the Brazilian's reception of the Aspen award in 1967.

In his own country, Gilberto Freyre occupies a position of influence and distinction rarely if ever attained in the United States by a historian or by a scholar of any sort. His American publisher considers him perhaps Brazil's "most distinguished living citizen." Frank Tannenbaum was of the opinion that *"The Masters and the Slaves* is a great deal more than just a book — it marks the closing of one epoch and the beginning of another" in Brazilian history. Freyre, he says, has worked a revolution in "Brazil's image of itself." Tannenbaum goes further to say that what cost "a bloody revolution, untold suffering, and the loss of a million lives" in Mexico was in Brazil "accomplished by one man and one book."[9] It is clear that we are dealing here with more than a phenomenon of scholarship.

No historian of the South should neglect to acknowledge his

[7] Gilberto Freyre, *The Masters and the Slaves: A Study in the Development of Brazilian Civilization,* second ed., (New York, 1956); *The Mansions and the Shanties: The Making of Modern Brazil* (New York, 1963). Freyre's *Order and Progress: Brazil from Monarchy to Republic* (New York, 1970) has less to say on slavery and race relations.

[8] W. R. Brock, an essay review in *History* (February 1967), 49.

[9] Frank Tannenbaum, Introduction to Freyre, *Mansions and Shanties,* vii, xi, xii.

obligation to Freyre for the wealth of insights his work has opened up for the study of American slavery, plantation life, patriarchal authority, and the cultural symbiosis produced by white masters and black slaves living together through centuries of time. More than any other scholar, the Brazilian historian has forced historians of the Old South to break out of their culture-bound parochialism and their race-limited outlook. The exceptions that will be taken below to some of the similarities Freyre finds and comparisons he draws between Brazilian and Southern slave culture are not to be interpreted as denying all similarities. Nor are they intended to dismiss his insight that slavery tended to pull all societies that practiced it in a common direction. To appreciate the extent as well as the limitations of his contribution, one must take into account some traits of their author.

Gilberto Freyre is a dedicated nationalist, or, as he would prefer, a "regionalist," who delights in the Brazilian way of life and celebrates things Brazilian. He is not indiscriminate or entirely uncritical in his celebration, but even those things he deplores he dwells upon with the relish of a connoisseur. And he dwells upon all strands in the colorful fabric of his tropical culture. Slavery and race relations are the focus of his subject, but they are the warp and woof of Brazilian civilization. It would be a great mistake to think of him as an apologist for slavery, though he does repeatedly make claims that slavery was less cruel, more humane, and more benevolent in Brazil than it was elsewhere, including French and Spanish as well as English-speaking countries.[10] Such comparisons and invidious claims about slavery are not, however, central to his interests or essential to his thesis, and he does not press them.

On the other hand, it would be correct to think of Freyre as a defender of miscegenation and to regard that defense of central importance in his work. He is very clear about this. After attending a Protestant missionary school at Recife, he came to the

[10] Freyre, *Masters and Slaves,* 183; *Mansions and Shanties,* 329; and *New World in the Tropics: The Culture of Modern Brazil* (New York, 1963), 198.

United States for a degree at Baylor University in Texas. "Every student of the patriarchal regime and the economy of slaveholding Brazil," he once wrote, "ought to become acquainted with the 'deep South.' " He was fascinated by the similarities and differences. But there and at Columbia University, where he took a Master's degree under Franz Boas, the youth with the Portuguese accent was depressed by talk he heard about "mongrelization." "I do not believe," he wrote, "that any Russian student among the romantics of the nineteenth century was more intensely preoccupied with the destiny of Russia than I was with that of Brazil." Of all its problems, "none gave me so much anxiety as that of miscegenation." Once on Brooklyn Bridge he encountered "a group of Brazilian seamen — mulattoes and *cafusos.*" He was overwhelmed with the feeling that they represented "caricatures of men, and then came to mind a phrase from a book on Brazil written by an American traveler: 'the fearful mongrel aspect of the population.' " [11] But from Boas he derived insights that liberated him from shame and he returned home to spread the joyful liberation among his beloved countrymen. This is what Tannenbaum meant about Freyre "changing Brazil's image of itself." After Freyre, Brazilians "no longer described themselves . . . as a mongrel race, inferior because it consists of a mixed people. On the contrary, they find their creative freedom, their pride in the present, their confidence in the future precisely in this fact — that they are a mixed, a universal people." Brazilians "turned their eyes inward and began to sing a song of themselves." They had "discovered themselves" and found "a ceaseless inspiration in being Brazilan and telling the world about it." [12]

The young historian, anthropologist, sociologist, folklorist (he rightfully claimed all these roles) had found his jewel in the toad's head of national shame. Miscegenation was his theme. It was the secret of the past and the wave of the future. It explained the greatness of Egypt, of Greece, of Rome — and of

[11] Freyre, *Masters and Slaves,* xxvi–xxvii.
[12] Tannenbaum, Introduction to *Mansions and Shanties,* xi.

Brazil.[13] It was the solution to the race problem and the path to ethnic democracy and social felicity. Anything that promoted it and facilitated its acceptance was of supreme interest to him. The fact that Freyre himself was a member of the white minority not only set up inner tensions but liberated him for the unrestrained candor with which he treated the subject.

The Portuguese, like the Puritans, had an "errand into the wilderness," but it was an errand of a different sort. "To the wilderness," writes Freyre, "so underpopulated, with a bare sprinkling of whites, came these oversexed ones, there to give extraordinarily free rein to their passions." The historian frankly admires their procreative energies. "Unbridled stallions is what they were," he says. And their opportunities were equal to their energies. "The milieu in which Brazilian life began," he says, "was one of sexual intoxication. No sooner had the European leaped ashore than he found his feet slipping among the naked Indian women," and even "many of the clergy" permitted themselves "to sink into the carnal mire." Miscegenation began and flourished as a Luso-Indian phenomenon that preceded and for a long time surpassed Luso-African activity. Not only was there a great shortage of European women, but because of the demand for men in Portuguese India, an appallingly short supply of Portuguese manpower for the colonization of so expansive an area as Brazil. So it was that "the lustful inclinations of individuals" were seen "to "serve powerful reasons of State, by rapidly populating the new land with mestizo offspring." The Jesuit fathers intervened in so far as possible to regularize libertinism by the sacrament of marriage, but the work went forward with or without ceremony.[14]

Only later, and not much earlier than they began to swell the

[13] Freyre, *Mansions and Shanties,* 429. On the rejection of racial purity and white supremacy by Brazilian nationalists in the early years of the twentieth century, see Thomas E. Skidmore, "Brazilian Intellectuals and the Problem of Race, 1870–1930," Occasional Paper no. 6 (1969), Graduate Center for Latin American Studies, Vanderbilt University.

[14] Freyre, *Masters and Slaves,* 29, 85.

population of English colonies, did Africans figure largely in Brazilian miscegenation. Contrary to the general impression that "it was for the most part the African who communicated to the Brazilians his lubricity," according to Freyre it was rather the other way around, "the Portuguese being the most libidinous" and the African the least among the three ethnic groups involved. Freyre is especially defensive about the reputation of the Negro woman in the whole commerce. It was not she "who brought to Brazil that vicious lustfulness in which we all feel ourselves ensnared . . . that precious voluptuousness, that hunger for women, which at the age of thirteen or fourteen makes every Brazilian a Don Juan." And "it was not the Negro woman who was responsible for this; it was the woman slave," a being at the service of an "idle master's economic interest and voluptuous pleasure. It was not the 'inferior race' that was the source of corruption, but the abuse of one race by another." [15]

However largely church and state figured in relations between master and slave, man and woman, or between dominant and subordinate races in Latin America, in Brazil the patriarch of the Big House, not viceroys and bishops, held absolute and virtually undisputed sway:

The Big House in Brazil, in the impulse that it manifested from the very start to be the mistress of the Land, overcame the church. It overcame the Jesuit as well, leaving the lord of the manor as almost the sole dominating figure in the colony, the true lord of Brazil, or nearer to being than either the viceroys or the bishops.

For power came to be concentrated in the hands of these country squires. They were the lords of the earth and of men. The lords of women, also. Their houses were the expression of an enormous feudal might. 'Ugly and strong.' Thick walls; deep foundations . . . enormous kitchens; vast dining rooms; numerous rooms for the sons and guests; a chapel; annexes for the accommodation of married sons; small chambers . . . Brazilian as a jungle plant. It had a soul. It was a sincere expression of the needs, interests, and the

[15] *Ibid.*, 94–95, 323, 329.

broad rhythm of a patriarchal life rendered possible by the income from sugar and the efficient labor of Negro slaves.[16]

"In Brazil," we are told, "the place of the cathedral or church, more powerful than the king, was taken by the plantation Big House." The Church was present, of course, and the Jesuits contested the planters' authority, but the clergy with whom the Big House dealt "grew big-bellied and soft in fulfilling the function of chaplains, ecclesiastical tutors, priestly uncles, and godparents to the young ones, and they proceed to accommodate themselves to the comfortable situation of members of the family or household, becoming allies and adherents of the patriarchal system." A Jesuit, says Freyre, saw "a danger that the padres would become subservient to the lords of the manor, and the perils, also, of too much contact — he does not state this clearly, but hints at it — with Negro women and mulatto girls." It is, in fact, hard to see how the Protestant clergy of Virginia and the Carolinas, with their ideas about sin, could have been more compliant with the planter regime and the slavery system. This tropical Catholicism was a "mild brand of household religion emanating from the chapels of the Big Houses" that were hung with images of "humanly friendly male and female saints." [17]

The patriarchal halls swarmed with "children of all ages and all colors eating and playing within the Big House, and they were all treated as affectionately as if they had been members of the family." And in shared paternity, they most likely were. "Bastards, natural sons — what sugar planter did not have them in large numbers?" asks Freyre.[18] In his opinion, "Our patriarchal ancestors were nearly always great procreators, and sometimes terrible satyrs, with a scapular of Our Lady dangling over their hairy bosoms. . . ." [19] But all living creatures of whatever legitimacy or origin found a place in the patriarchal order. "The hierarchy of the Big Houses was extended even to

[16] *Ibid.*, xxxv, xli. [17] *Ibid.*, 169, 192, 372. [18] *Ibid.*, 399, 437.
[19] *Ibid.*, 379.

parrots and monkeys," which received "the benediction from Negro lads, just as the lads received it from the aged blacks, who in turn were blessed by their white masters." [20] In retrospect the Big House-Slave Quarters system seems to the Brazilian historian "in certain respects, a veritable marvel of adjustment: of slave to master, of black to white, of son to father, of wife to husband." [21]

The adjustment of the planter's legitimate sons was facilitated by his black slave playmate, "an obliging puppet, manipulated at will by the infant son of the family, he was squeezed, mistreated, tormented just as if he had been made of sawdust. . . ." This probably encouraged the white son "in sadistic and bestial forms of sexuality. The first victims were the slave lads and domestic animals; but later come that great mire of flesh: the Negro or mulatto woman." The fathers, "dominated by the economic interest of the masters of slaves, looked always with an indulgent and even a sympathetic eye upon the precocity of their sons in anticipating the generative functions and . . . they even made it easy for these young stallions to do so. Rural traditions tell of mothers, untrammeled by morals of the case, who would thrust into the arms of their bashful and virginal sons young Negro and mulatto girls who were capable of awakening them from their apparent coldness or sexual indifference." Typically, we learn, the son "lost no time in taking Negro women that he might increase the herd and the paternal capital." This picture was "at times marked by an unpleasant trace of incest," since the son's sire had normally engaged in the same domestic activity before him and had often left daughters in the slave huts.[22]

Precocity was demanded of sons in other respects as well, for they were "compelled to become little men from the age of nine or ten years. . . . Their garments were those of men. Their vices those of men. Their concern: to syphilize themselves as soon as possible, thereby acquiring those glorious scars in the bouts of Venus." Indeed the lad of thirteen "would be the butt

[20] *Ibid.*, xlii. [21] Freyre, *Mansions and Shanties,* xxiv.
[22] Freyre, *Masters and Slaves,* 349, 395, 356.

of jests if he could not show the scars of syphilis on his body." Freyre is especially concerned with the mistake foreigners make in "attributing to miscegenation effects that are chiefly due to syphilis"—physical afflictions and deformations widespread among mixed breeds. He points out that syphilis "was, *par excellence,* the disease of the Big House and the *senzalas,"* and that "Brazil would appear to have been syphilized before it was civilized," instead of vice versa. It was regarded as "a domestic disease, a family one, something like measles or the worms" and accepted fatalistically. "It was a serpent brought up in the house, with no one taking any notice of its venom." [23]

Daughters of the patriarchal Big House were "denied everything that savored in the least of independence," and "lived under the stern tyranny of their fathers — and later under the tyranny of their husbands." [24] As Freyre observes, "We Brazilians liberated ourselves more quickly from racial prejudice than from those of sex. . . . 'The inferiority of women' took the place of 'the inferiority of race.' " [25] Precocity in sexual life was demanded of daughters as well as of sons, but with striking limitations. "They married at twelve, thirteen, or fourteen years" and the grooms were "thirty or forty, and at times of fifty, sixty, or even seventy." By the time a girl was fifteen she was, if not married, "a pitiable old maid already." White women aged so quickly that "their charms barely lasted until the age of fifteen" and they declined into "little old ladies of thirty or forty years." Negro and mulatto women retained their bloom much longer.[26]

The premature fading of white women is attributed not only to early and frequent childbearing but to "the morose, melancholy, indolent life they led inside the home." [27] Virtual prisoners of the Big House, they were not permitted to see the visitors of their husbands or to dine in their company. Confined to the company of slaves, they sometimes grew up and remained illiterate and absorbed African voodoo and endless lore of love po-

[23] *Ibid.,* 404, 70–71, 326–327. [24] *Ibid.,* 418.
[25] Freyre, *Mansions and Shanties,* 99–100.
[26] Freyre, *Masters and Slaves,* 361–364, 379. [27] *Ibid.,* 360.

tions and charms. It was an atmosphere festering with "rivalry of woman with woman" and rent with outbursts of rage against "scheming slave girls" and revenges of a sadistic sort. Freyre relates "many instances of the cruelty of ladies of the Big House toward their helpless blacks" — that of ladies "who had the eyes of pretty mucamas gouged out and then had them served to their husbands for dessert, in a jelly dish," and of others who kicked out the teeth of their women slaves with their boots, or had their breasts cut off, their nails drawn, or their faces and ears burned. A whole series of tortures." [28] While he considers the sadistic tendency more characteristic of the distaff side, Freyre tells of one patriarch who "upon discovering that his son was having relations with a favorite slave girl, had him slain by an older brother," and of others "having pregnant slave girls burned alive in the plantation ovens, the unborn offspring crackling in the heat of the flames." [29]

These are not the aspects of patriarchal life that enshrined the Big House in Gilberto Freyre's heart, and it is much to his credit that he is candid enough to admit their presence. He also admits that a neurotic gloominess and melancholy often pervaded the massive halls and that they could be veritable "brothels" and "dens of perdition." For all that, he persists in his faith that "it was in the Big House, that, down to this day, the Brazilian character has found its best expression." It is as if some dedicated American abolitionist such as James G. Birney or Angelina Grimké could, along with their searing exposures of slavery, have appended to their indictments apostrophes to the Plantation Legend — an improbable blend of an antislavery Theodore Weld and a proslavery George Fitzhugh. But it is the Fitzhugh in Freyre that predominates — the defender of patriarchal values and their plantation seat. "It had a soul."

"First of all," writes the Brazilian, "I would emphasize the prevailing mildness of the relations between masters and household slaves — milder in Brazil, it may be, than any other part of

[28] *Ibid.*, 351. [29] *Ibid.*, xxix, xliii, xlvi.

the Americas."[30] Though in the Portuguese titles of his two main works "Big House" is coupled with "slave hut" and "mansions" with "shanties," he has little to say about life in the latter. Indeed, there is little at all about the great mass of slaves, the field hands of the sugar and coffee plantations. It is the house slaves he is concerned with, their relations with the masters and whites: the communal life of Big House. He can be as brusque as George Fitzhugh about the "antislavery sentimentalism or the doctrinal zeal of those who try to fit the history of patriarchal societies into this or that 'ism,' that in such societies the slave was always and from every point of view a 'martyr,' a 'victim,' 'undernourished.' The truth of the matter is that there were societies like Brazil, in which, by and large, the slave of the authentically patriarchal regions was better nourished, better treated, led a better life than those already industrialized or commercialized."[31]

When he refers in passing to the coffee plantations "down south," Freyre can be as harsh as any doctrinaire about "martyrs," "victims," and "undernourishment" — as harsh as Fitzhugh on the dark, satanic mills of Massachusetts and their "slaves without masters." For the coffee boom represents to Freyre "the transition from the patriarchal to the industrialized economy, with the slave less a member of the family than a mere worker or 'moneymaking machine.' " It was "a system no longer patriarchal but perverted by a rapid . . . imitation of bourgeois commercial industrialism." And where he writes of "the horror of the slaves of the pleasantly patriarchal northeast," his native region, "when their masters threatened on the days of their blackest anger that they would sell them to the coffee plantations of São Paulo," he echoes a familiar theme of the Virginia or Maryland Tidewater gentry unable to take care of "their peo-

[30] *Ibid.*, 369. For an astute comparison that comes to an opposite conclusion see Carl N. Degler, "Slavery in Brazil and the United States: An Essay in Comparative History," *American Historical Review*, LXXV (April 1970), 1004–1028. [31] Freyre, *Mansions and Shanties*, 186.

ple" and forced to "sell them down the river" — to sugar planters of Louisiana or cotton planters of Mississippi.[32]

The down-the-river gradation of North American slavery horrors, however, does not seem to include a level comparable to that reached on the coffee plantations and the mines of southern Brazil. There were individuals, whole plantations, even groups of plantations in the Old South that earned and probably deserved reputations for ruthlessness and cruelty, but not whole regions and industries that earned for long periods a name for comparable evil. Freyre's estimate of the dehumanization and alienation of the slaves in southern Brazil is not only sustained but extended and darkened by more recent and more thorough scholarship.[33] But what of the slaves in "the pleasantly patriarchal northeast," his native region? Freyre gives scant attention to those outside the household, but such generalizations as he does make about them draw a picture of contentment and paternal benevolence. He finds persuasive evidence that slaves under the paternal regime "were more or less resigned to their status. The exceptions seem to have been rare." A slave quoted by a traveler in 1830 as saying "that he was satisfied with his life as a slave" is presented as "representative or typical of the slaves of his epoch, that is to say, those who received fatherly treatment at the hands of their masters." [34] The picture recalls those often drawn of the mellowed patriarchies of the Chesapeake area after the decline of the tobacco boom. Freyre does not explain how patriarchal benevolence extended over vast estates much larger than any in the United States, with more than two thousand slaves, and how prevalent it was in all centuries and stages in the long history of the sugar industry. There were periods, especially during the Empire, of which relaxed benevo-

[32] *Ibid.*, 130–131, 186–187.

[33] Stanley Stein, *Vassouras, A Brazilian Coffee Country, 1850–1900* (Cambridge, Mass., 1957); C. R. Boxer, *Race Relations in the Portuguese Colonial Empire, 1415–1825* (Oxford, 1963); Degler, "Slavery in Brazil and the United States," 1019–1028.

[34] Freyre, *Mansions and Shanties*, 330–331.

lence and mildness were characteristic, just as they characterized certain periods and areas of American slavery. Sternness and harsh discipline are not inconsistent with paternalism, may indeed be characteristic of it. Freyre's own accounts of sadistic floggings (*novenas* lasted nine to thirteen consecutive nights), fiendish tortures, and bloody, massive, and sustained slave insurrections involving thousands and lasting decades — not to mention grimmer details supplied by other scholars — belie extension of the benevolent description beyond limited phases.[35] Making full allowance for the status of household slaves, C. R. Boxer holds that "it remains true that by and large colonial Brazil was indeed 'a hell for blacks.' "[36]

But in fairness to Freyre, it is the household slave about whom he is mainly thinking and on whom he rests his case, the slaves in and around the Big House of the plantation and the mansions of the cities. "It was the Negro who gave to household life in Brazil its cheerful note. . . . It was the Negro's hearty laugh that broke in upon the 'dull, abject mournfulness' that tended to stifle the life of the Big House." The Negro, with a tolerant church presiding, "filled the popular feast-days of Brazil with reminiscences of his own totemistic and phallic cults." Negroes were "all the while singing and dancing, exuberantly." They "always sang as they labored," and their work songs as well as "their feast-day melodies, and their cradle songs filled Brazilian life with an African joyfulness."[37]

Many of the scenes, images, roles, and stock figures of Big House life are interchangeable with their counterparts of the Plantation Legend in the Northern hemisphere. The master would send for Negroes to sing their work songs and religious songs "whenever a visitor arrived." Black mammy evoked "a

[35] Davis, *The Problem of Slavery* (cited in note 4), 233–241; Arthur Ramos, *The Negro in Brazil* (Washington, D.C., 1939), 20–23; Donald Pierson, *Negroes in Brazil: A Study of Race Contact in Bahia* (Chicago, 1942), 47–48.

[36] Boxer, *Race Relations* (cited in note 6), 114.

[37] Freyre, *Masters and Slaves,* 472.

depth of tenderness of which Europeans do not know the like." Privileged black characters "were given their way in everything." Saintly black retainers who were "capable of transmitting to white children a Catholicism as pure as that . . . received from their own mothers" recall those Cindys, and Dilceys and Ellas — "the best Christians of us all, I guess" — from the Tidewater to the Delta. And those eulogies of "the purity of Brazilian women" like those to "Southern womanhood," evoke the familiar reflection that "much of this purity rested upon slave prostitution." Slave huts back of the Big House might be built of straw, but "the sun comes into them like a rich generous friend of the family." The long experience of slavery left the imprint on the personality structure of the Negro and his descendants, as it did in North America, but it is the sunny side of "Sambo" that appeared uppermost — just as it did below the Potomac. For the slave system of Brazil "would seem to have developed in the slave, and in his mulatto descendant, agreeable manners which came from the desire of the slave to win the good graces, if not the love, of the masters." The agreeable manners were graced by "a gentle manner of speaking, accompanied by a pleasant smile." [38]

Freyre's references to the American South are frequent and almost invariably friendly. He delights in evidence that their common institutions pulled Brazil and the South together. It seems apparent that he retained fond memories of his college days in provincial Texas and associated much that he saw in the South in a youthfully nostalgic way with his beloved homeland. He formed friendships with Southerners, revisited the South in later life, and is fond of comparing cookery, manners, and mores of the two postslavery cultures. The comparisons, historic as well as contemporary, are usually genial if not flattering — always with the reservation about the South's rigidity of racial attitudes. But his writings would not seem motivated by the invidious uses to which they have sometimes been put by modern critics of the American South. He is more interested in the similarities than in

[38] *Ibid.*, 369, 370, 372, 455, 472; *Mansions and Shanties*, 381, 406.

the differences between the Old North of Brazil and the Old South of the United States, and if anything he is prone to exaggerate them:

In the South of the United States, there evolved, from the seventeenth to the eighteenth century, an aristocratic type of rural family that bore a greater resemblance to the type of family in northern Brazil before abolition than it did to the Puritan bourgeoisie. . . . There were almost the same country gentlemen — chivalrous after their fashion; proud of their slaves and lands, with sons and Negroes multiplying about them; regaling themselves with the love of mulattos; playing cards and amusing themselves with cockfights; marrying girls of sixteen; engaging in feuds over questions of land; dying in duels for the sake of a woman and getting drunk at great family feasts — huge turkeys with rice, roasted by "old mammies." [39]

One fabulous but elusive, and hitherto wholly neglected opportunity for comparison of the Protestant, Anglo-American Old South system with the Catholic, Luso-American tropical society, suddenly placed in physical juxtaposition, came hard on the downfall of the Lost Cause in the last two decades of Brazilian slavery. During the three years following the defeat of the Confederacy, eight to ten thousand voluntary Confederate exiles sought to establish themselves in various Latin American countries. The majority of them settled in Brazil. In the years just preceding the Civil War, Matthew Fontaine Maury, the famous Virginia scientist, had pictured the Amazon Valley as "The Garden of the Hesperides," a marvelous tropical counterpart to the valley of the Mississippi capable of accommodating a hundred million settlers. Actually only a few of the Confederate exiles settled on the Amazon. In 1867, a colony of two hundred, mainly the families of Alabama and Tennessee planters, settled five hundred miles from the mouth of the river. A few prospered and remained, but most of them gave up and returned. The great majority sought more temperate parts (with Rio de Janeiro as

[39] Freyre, *Masters and Slaves*, 401.

the center of activity), bought slaves and plantations, built Protestant churches, and many dug in to stay. The exiles were not unaware of the frying-pan-to-fire irony of their choice of refuge in view of their motives. Having fled as intolerable the Radical Reconstruction in their native land and its innovations of racial equality, they landed in a country where free blacks enjoyed "even more rights than the white foreigner." How they made their peace with the new social and cultural environment, were it known in depth, would likely make one of the most fascinating chapters in historical sociology. But an observant visitor to the Brazilian grandchildren of the Confederate exiles in the 1950's found all about him haunting echoes of the accents, attitudes, and racial relations of the gentler aspects and tones of rural black-belt Alabama.[40]

Freyre overlooks the Confederate exile opportunity. In his comparisons between the Brazilian Old North and the American Old South, he dwells on such phenomena as laziness, hospitality, extravagance, conspicuous waste, soil mining, horse racing, hunting, cardplaying, swearing, excessive individualism, "a marked fondness for long and noisy conversations," and a readiness to confer the rank of "colonel." He finds equally true of Brazil a remark Ulrich B. Phillips made about the singular freedom of the plantation system of the Old South from "the curse of the impersonality and indifference which too commonly prevails in factories of the present-day world where power-driven machinery sets the pace, where the employers have no relations with the employed outside of work hours." [41] Freyre does note two significant differences between Brazilian Old North and American Old South plantation orders. One is what appears to him to be an inferiority of American Negroes to those of Brazil, which got "the best that African Negro culture had to offer,"

[40] Lawrence F. Hall, "The Confederate Exodus to Latin America," *Southwestern Historical Quarterly*, XXXIX (1935–36), 100–134, 161–199; also Hamilton Basso, "The Last Confederate," *New Yorker* (November 21, 1953), 143–161.

[41] Freyre, *New World in the Tropics*, 82–92; *Mansions and Shanties*, 183.

"the elite elements," and "the top soil of black people," while these elements were "lacking in the same proportions in the Southern United States." The second was that in Brazil "the family was to be enlarged by a far greater number of bastards and dependents, gathered around the patriarchs, who were more given to women and possibly a little more loose in their social code than the North Americans were." He also concedes that American women were less tolerant of husbandly promiscuity and tyranny. But apart from these differences and the Anglo-American rigidity about racial purity, the two cultures seemed to him at times "so similar that the only variants to be found are in the accessory features: the differences of language, race, and forms of religion." [42]

To a Protestant Anglo-American of the temperate zone, however, the "accessory features" might seem of somewhat more than incidental importance. The culture contrast suggests setting a flock of gray and white mockingbirds down in a tropical jungle filled with gaudy parakeets. To Puritan New England, life along the James, the Sewannee, or the Lower Mississippi may have appeared lushly exotic and outlandishly bizarre. But set side-by-side with life along the Amazon, the colors of antebellum society in the Old South fade to temperate-zone grays and russets and muted saffrons that went well enough with magnolias or Spanish moss, but were not quite the thing for promenades under palm and breadfruit. Beside the garish social scene in the tropics, the Old South was a subdued affair of demure ladies in crinoline and pantalettes, gentlemen in black stock and broadcloth, and sweaty Sunday mornings in Episcopal pews or plain white Methodist chapels. Gaiety there was and spontaneity too, along with some luxury and display, more than a dash of wicked exuberance, and a pervading undertone of more or less repressed sexuality. But it was all under an umbrella of Protestant, Anglo-Saxon restraint and inhibition.[43]

[42] Freyre, *Masters and Slaves,* 299, 306, 311.

[43] Of the many illustrations that come to mind, perhaps the readiest is Mary Boykin Chesnut, *A Diary from Dixie* (New York, 1949).

The historic achievement of Brazil, which in the eyes of the patriot historian places its civilization among the greatest of all ages, is its ideal adaptation of European culture and technology to "tropical situations." He proclaims this the birth of a new science and gives it the formidable name of "Lusotropicology." It seems that the Portuguese "have been expanding Lusotropically and Christocentrically" until in Brazil they have built "one of the most socially, culturally and technically advanced countries in the world." The secret of the new science lies in felicitous combination of things European with things tropical — not only between races (by the gift of "Lusomiscibility") but with food, clothing, housing, dancing, and religion, indeed all things human. One recommended product of "Lusotropical symbiosis" is "the combination of Scottish whisky and coconut milk." If this particular combination, and perhaps some others, sets the teeth of temperate-zone folk on edge, they might consider Freyre's warning to his own people against "fashions and norms of cold or temperate countries, valid only for those countries." For if tropical folk err in "going against tropical ecology," and if it is indeed "absurd for the tropical people to resign themselves to passively adopt from such cold or temperate civilizations everything," then perhaps temperate-zone people might reasonably be expected to adopt the same selectivity and restraint about tropical inventions.[44]

It is not only the tropical variable that complicates the Brazilian comparison, but also "the Oriental influence on the life, landscape, and culture of Brazil." Freyre makes much of the degree to which Brazilian culture was "impregnated with Moorish, Arabic, Israelite, Mohammedan influences," mainly through its Iberian heritage and Portugal's oriental colonies. He offers as illustrations the palanquin, the litter, lattice windows, "brightly painted houses in the form of a pagoda," upturned eaves, long

[44] Gilberto Freyre, *Portuguese Integration in the Tropics* (Lisbon, 1961), 24, 49, 52, 58, 61; *New World in the Tropics,* 22, 24, 26, 142; *The Portuguese and the Tropics* (Lisbon, 1961), 62.

fingernails ("a hideous length" affected by both sexes), women's shawls and turbans, the unexampled use of the perfumes of the East ("to cover up the so-called Negro smell") and hundreds of others. The use of sedan chairs or palanquins was widespread as late as the nineteenth century. "They were trimmed with gold and silver, and hung with heavy curtains, sometimes of silk, and adorned with figures of cupids, angels, dragons." Those of the wealthy were "carried by Negroes in colored livery-frock coats, breeches, blue and red kilts, even though they went barefoot. . . ." [45] As an exercise in comparative history, one might try fitting this conveyance into the cultural landscape of Virginia and the Carolinas — or even Mississippi in the flush time of the cotton nabobs. Not even King Carter in his pea-green coach-and-six with outriders could hold a light to this spectacle.

There is a certain oriental flavor, too, about the description of the life of the Brazilian planter. Slaves "became literally their masters' hands — at least, their right hands; for they it was who dressed them, bathed and brushed them, and hunted their persons for fleas. . . . The Master's hand served only for telling the beads of Our Lady's rosary, for playing cards, for taking snuff . . . and for playing with the breasts of young Negro and mulatto girls, the pretty slaves of his harem. In the case of the slave-owner the body became little more than a *membrum virile*. . . . Slothful but filled to overflowing with sexual concerns, the life of the sugar-planter tended to become a life that was lived in a hammock. A stationary hammock, with the master taking his ease, sleeping, dozing. Or a hammock on the move, with the master on a journey or a promenade beneath the heavy draperies or curtains. Or again, a squeaking hammock, with the master copulating in it. . . . The life of the sugar-raising aristocrats was a soft and languid one. . . . Ordinarily the days went by, one like another, with the same lethargy, the same uneventful and sensual hammock existence." [46]

[45] Freyre, *Mansions and Shanties,* 273–288.
[46] Freyre, *Masters and Slaves,* 428, 430.

Critics say this is not really fair to Brazilian planters. That is not the question. The point is that at this juncture the comparative impulse seizes upon their historian and he is impelled to write that the same economic institutions, "prevailing over the differences in climate, race, and religious morality, created in the Southern United States a type of frail, soft-handed aristocrat who was scarcely to be distinguished from the Brazilian type in his vices, his tastes, and even his physique. The ingredients were different, but the form was the same." [47]

The picture of the sugar planter of Brazil is derived from a friendly or fondly indulgent source. But not even the inflamed imagination of the most hostile American abolitionists produced a serious characterization of the Virginia tobacco planter, the Georgia cotton planter, or the Louisiana sugar planter that bears a resemblance to this hammockful of oriental sloth, vermin, and sensuality. This is not to insist that the latter was a typical adaptation of "Lusotropical ecology," nor to risk the smiles of anthropologists for a lapse into ethnocentric prudery or culture-bound naïveté. It is only to point out that ecology and morals undergo a sea change in the horse latitudes and that while the resulting adaptations varied a good deal in planter types from seventeenth-century Tidewater tobacco to nineteenth-century Delta sugar, they did not typically tend to be soft-handed, supine bodies reduced to the *membrum virile*. Any such adaptation would have precipitated bankruptcy much quicker than it normally came. If the Puritan Ethic below the Potomac subsides, compared with New England, it bristles stridently compared with Brazil. More typically, the planter-aristocrat of these latitudes was a hard-handed, hard-riding, hard-living patriarch, alert to a thousand demands and decisions of his own estate, while he undertook on the side to run state, church, and society — and sometimes the nation as well.

Racial attitudes and policies of Old South society were rigid, often unjust, too often inhumane, and they undoubtedly left a heritage that handicapped the adjustment of freedmen to free-

[47] *Ibid.*, 430.

dom and the races to each other. It might have been better to have permitted the Africans to retain much of their own culture and for the churches, like the indulgent priesthood of Brazil, to tolerate the intrusion of African divinities among the saints and voodoo rituals into the litany. Protestant slaves might thus have been under less compulsion to adopt white gods and white world views. Here again, stern anthropological variables of a Protestant culture obtruded like primeval rock to block the path of integration. There was no place on bare white-paneled walls of Methodist churches for black Madonnas — or Madonnas of any color. The tolerance of Baptist congregations for African totemistic and phallic cults was somewhat limited. And the possibilities of turning a protracted camp meeting or an evangelical revival into an Afro-Roman fiesta with dinner on the ground were not very encouraging.[48]

In tropical Brazil, of course, things were different. There the path to racial integration "was smoothed by the lubricating oil of a deep-going miscegenation." But miscegenation also flourished, if rather less luxuriantly, in the Old South as well — only the "lubricating oil" did not serve the same social ends. As Freyre remarks, "There is no slavery without sexual depravity." [49] The incontrovertible evidence of this walked in and out of some of the finest mansions of the Old South, but was rarely mentioned in polite circles.

In one of the unpublished passages of her manuscript diary, Mary Boykin Chesnut quotes a planter friend in Montgomery in March 1861 as remarking in praise of the Southern way of life that "it was so patriarchal." She adds: "So it is — flocks and herds of slaves — and wife Leah does not suffice — Rachel must be added — if not [illegible word] and all the time they seem to think themselves patterns — models of husbands and fathers." [50] Then three months later, after the Civil War had

[48] Melville Herskovits, *The Myth of the Negro Past* (Boston, 1958), 72, 249. [49] Freyre, *Masters and Slaves,* 182, 324.

[50] Manuscript diaries and journals of Mary Boykin Chesnut (Caroliniana Library, University of South Carolina, Columbia, S.C.). This is

started, back at Mulberry Plantation, her lovely home in South Carolina, she enters (again, unpublished) a muffled cry of anguish: "Merciful God! Forgive me if I fail. Can I respect what is not respectable? Can I honor what is dishonorable? *Rachel* and her brood makes this place a horrid nightmare to me. I believe in nothing with this before me." [51] And in one of the published editions of the *Diary* she declares, "Under slavery, we live surrounded by prostitutes. . . . God forgive us, but ours is a monstrous system, a wrong and an iniquity! Like the patriarchs of old, our men live all in one house with their wives and their concubines; and the mulattoes one sees in every family partly resemble the white children. Any lady is ready to tell you who is the father of all the mulatto children in everybody's household but her own. Those, she seems to think, drop from the clouds. My disgust sometimes is boiling over. Thank God for my countrywomen, but alas for the men! They are probably no worse than men everywhere, but the lower their mistresses, the more degraded they must be." [52]

Perhaps it is one consequence of Anglo-Saxon reserve or inhibition, but we have not yet had — indeed, may never have — a historian of Southern society who combines the candor and the sexual curiosity that Gilberto Freyre brought to his work on Brazilian slavery and race relations.[53] Much remains to be known, and much is buried with those poker-faced ladies of the stiff oil portraits and tintypes that will never be known about life in the Big House and the Slave Quarters below the Potomac. There are only hints, suppressed passages of diaries, bundles of

---

one of the last entries under March, 1861, immediately following the passage in the published version (Chesnut, *Diary from Dixie,* 22) quoted at the end of this paragraph.

[51] Manuscript diaries and journals of Mary Boykin Chesnut. This passage was omitted from entries published on p. 58 of her *Diary from Dixie.* [52] Chesnut, *Diary from Dixie,* 21–22.

[53] Meanwhile, a black historian has made a substantial contribution of the sort: James Hugo Johnston, *Race Relations in Virginia* and *Miscegenation in the South, 1776–1860* (Amherst, Mass., 1970).

old letters, an occasional revelation in a will or in court. One can only speculate.

It is plain enough that the ladies in crinoline had a lot to put up with. And they did put up with it, many of them, but only in their own fashion. Their fashion raises some interesting questions of comparative candor and morals of the sort that lace all contrasts between the Catholic-Latin-tropical culture and the Protestant-Anglo-Saxon-temperate one. The ladies were reticent, evasive, often willfully blind about what sometimes went on in their own backyards. But what was shameful was regarded as shameful. It was not condoned as the legitimate prerogative of patriarchs, the proper initiation to manhood for one's sons, or an acceptable means of increasing one's labor supply. Nor was it brought into the parlor and flaunted in the streets. Nor was it decked out as a polygamous extension of patriarchal benevolence. Except for a period in the history of Utah, harems did not flourish openly in the Protestant-temperate zone. As for the men, they did what they did, but they were careful what they said about it around the house.

As for the ladies in crinoline, their grandmothers in silks, or their daughters in bustles, it is impossible to imagine them putting up with anything like the domestic tyranny, the oriental purdah, and the degrading humiliation that were the *senhora*'s daily lot. They may have sometimes married at sixteen or seventeen (not at twelve and thirteen), but when they did, they took over as adult women in full charge, and not as adolescent playthings of men three or four times their age. Nor were they about to acquiesce submissively in sharing a husband with a slave harem, not willingly at least, nor without pressures they knew best how to apply, and when they failed and knew about it at all, it was with the anguish of spirit that Mary Chesnut expressed with such bitterness, and not with a fatalistic acceptance of sanctioned mores.

More Protestant prudery, more Anglo-Saxon concern for appearances, more white supremacy? Perhaps. But the question is one of candor. And, after their fashion, the ladies in crinoline were calling a spade a spade. They knew without daring to articulate it that slavery meant, among other things, defenseless women at the disposal of men. Or, as they probably put it, "Men are beasts." And they assumed, rightly or wrongly, that sexual exploitation of slave women is shameful — in any culture. Besides, they had their pride. Too much pride, or too much candor, to give in to egregious male fictions about the natural needs of the sex. Too much pride to accept one of the favorites at their table or in their church pew. Such concessions might be made with impunity in other cultures, other climes. But not in theirs. So they set their chins, and they dug in their heels, and that was that.

Resolute wars for the defense of integrity, especially wars against great odds, come at a cost. Other values, often important ones, get sacrificed. This war was no exception. And some of the sacrificed values in this instance were basic ones of Christian charity, patriarchal responsibility for "one's people," decent regard for one's dependents, blood ties with one's kin, the tribal imperatives. Blood, sweat, and tears went into the fulfillment of paternalistic codes about the care and welfare of "the servants," sleepless nights ministering to distress in the quarters, and all that. But the fact remained that these people, blood kin along with them, were excluded, shut out resolutely from basic human ties, hopes, aspirations, opportunities. The ultimate horror might be the tragedy of Charles Bon in *Absalom, Absalom!* or that of Joe Christmas of *Light in August*. But the ultimate cost has never been reckoned. It is still unpaid, still mounting, and it could run much higher.

Perhaps it really would have been better had the Anglo-American masters, like those of Iberian origins, undergone the experience of seven centuries of rule by a people such as the Moors, who were of somewhat darker color and higher culture than the ruled and who practiced a civilized polygamy and harbored no aversion to miscegenation. Perhaps they should never have settled in a temperate zone and transplanted English institutions. Perhaps it would have been better, for that matter, never to have broken with Rome, never to have quarreled with monarchy, never to have established those proudly autonomous colonial assemblies, and never to have married such women as they did. But, then, that would have been somebody else's history and not our own.

3

Southern Slaves in the World of Thomas Malthus

Any historical wrong with a reputation as monstrous as that of the Atlantic slave trade puts a strain on the historian's vocabulary and his techniques of portrayal. Words, even the anguished words of the victims or the indignant words of the abolitionists, seem inadequate. There is a natural impulse to compensate with statistics, or at least to make numbers live up to the enormity of the crime. One result has been a tendency to exaggerate, sometimes wildly, that has persisted into recent times. Thus one reputable historian of the trade estimates that between 40 and 50 millions of Africans were exported to the West Indies in the period ending twenty years before the British made importations illegal. And "the most moderate estimates" of Brazilian historians were cited in 1935 as placing the number shipped to Brazil alone after the first century and a half of the trade at "no fewer than 30 million blacks." [1] These estimates represent the upper ranges of exaggeration. Much more common is a wide consensus that the total number of slaves imported into the Americas through the centuries came to between 15 and 25 million. Adopted by an impressive list of authorities and continued

[1] The first figure is attributed to P. Dieudonne Richon (1929), and the second to Benjamin Perit (1934), quoted in Robert R. Kuczynski, *Population Movements* (Oxford, 1936), 13.

down to the present day, this consensus proves to have been based on a pyramiding of insubstantial guesses and estimates that gained authority by sheer repetition.[2]

The share of these exaggerated slave imports assigned the United States and the original colonies has itself been subject to exaggeration for various reasons. For one thing, the enormous number of slaves emancipated in 1865, more than four million of them,[3] encouraged the assumption of huge imports to account for the huge population. No other contemporary country, colony, or empire ever approached that number of slaves at any one time. Brazil, the next largest, reached a maximum of slave population in 1850 with some 2,500,000.[4] The entire British Empire, including uncounted slaveholdings in the Eastern hemisphere, contained fewer than 800,000 slaves at the time of emancipation in 1834, while the slaves of the United States already exceeded two million by 1830. At any time between 1840 and 1865 it is probable that the United States held more slaves than all other slave powers combined.

The assumption that the size of slave population was a reliable indication of the number of slaves imported has resulted in enormously inflated estimates. Because the combined slave population of Jamaica and Saint Domingue roughly equaled the number of slaves in the United States in 1789, a modern historian concluded that the latter must have imported a number equivalent to the imports of the two Caribbean islands. He therefore estimated the North American imports by 1789 to have been 1,500,000, which were probably more than five times

[2] The historiographical evolution of this vast consensus is traced in a fascinating chapter entitled "The Slave Trade and the Numbers Game: A Review of the Literature," by Philip D. Curtin, *The Atlantic Slave Trade: A Census* (Madison, Wis., 1969), 3–13.

[3] The number of slaves listed by the census of 1860 was 3,953,760. The increase during the next five years is impossible to estimate, and any estimate is rendered meaningless by the effect of Union army invasions. But in the decade before 1860, the slave population had increased 23.4 percent.

[4] Stanley Stein, *Vassouras, A Brazilian Coffee Country, 1850–1900* (Cambridge, Mass., 1957), Appendix, 294.

the actual number.[5] Most historians have been much more cautious in their estimates of slaves imported in the colonial and early national era.[6]

The period after the close of the legal slave trade, from 1808 to 1860, however, has inspired less caution and greater exaggeration. The number of slave ships fitted out in American ports, the many unfounded rumors of slave landings, an unquestionable laxity of law enforcement, the Southern agitation for reopening the slave trade, and the suspicions bred by current agitation of the slavery issue all contributed to overestimates of illegal imports. These estimates range from a quarter of a million to a million.[7] Undoubtedly some slaves were landed in this period, but a close investigation has led one historian to the conclusion that "the smuggled slave remains so elusive that one may wonder whether in fact he ever existed in very large numbers." [8]

The long-overdue revision of slave trade statistics and history has at last been precipitated by a work of prime significance by Philip D. Curtin, *The Atlantic Slave Trade: A Census*. Building with existing data and printed sources, seeking to eliminate inconsistencies and exaggerations, to overcome national and ethnocentric limitations of previous work, and to see the story in the large, Curtin has attempted to determine "the measurable number of [African] people brought across the Atlantic. How many? When? From what parts of Africa? To what destination in the New World?" If readers seize upon his answers to these questions as definitive and regard his figures as exact, it is no fault of the author. He warns that this is "not intended to be a

[5] Noel Deerr, *The History of Sugar* (2 vols., London, 1949–50), II, 49, 72–75, 84–86, 284; Curtin, *Atlantic Slave Trade,* 87.

[6] On estimates of slave imports in the early period see Lewis C. Gray, *History of Agriculture in the Southern United States to 1860* (2 vols., Washington, 1933), II, 648–649.

[7] The estimate of one million is that of Deerr, *History of Sugar,* II, 282. Other estimates, including W. E. B. DuBois's figure of 250,000, are summarized in Gray, *History of Agriculture,* II, 649.

[8] Warren S. Howard, *American Slavers and the Federal Law, 1837–1860* (Berkeley, 1963), 154 and note 22, pp. 302–303.

definitive study," and admits that "the hoped-for range of accuracy [in most statistics] may be plus or minus 20 percent of actuality." He harbors no illusions that figures are "more scientific" than other kinds of data and offers his own "only as the most probable figures at the present state of knowledge." [9]

Even so, what he does offer are far and away the best informed estimates that we have. His book marks a turning point, not only in the study of the slave trade, but in the comparative history of slavery and the whole African experience in the New World. Scholars will be busy throughout the Americas for a long time to come applying, adding to, and correcting his findings. The purpose here is not to correct or add to his statistics, but the more modest one of exploring some of their implications for the history of slavery and the Afro-American in the South and testing new comparative perspectives that they suggest.

Most striking on first impression is the downward revision of the old consensus on the overall total of slaves imported into the Americas. Working with country-by-country estimates, Curtin arrives at a total of 9,566,000, which includes 175,000 for Old World imports by the Atlantic. He insists that the estimate "be understood to indicate a range of possibility, not a set figure . . . to be repeated in the textbooks." The available data are too full of holes and uncertainties to justify any more than that. Yet there is a greater probability that the estimate is too high rather than too low, and in any case "it is extremely unlikely that the ultimate total will turn out to be less than 8,000,000 or more than 10,500,000." The only competing estimate on comparable lines, that of Noel Deerr, is 25 percent higher. The difference is mainly accounted for by Deerr's huge miscalculation of slave imports for British North America and the United States, which he estimates to be 2,920,000. Eliminating that difference, Deerr's estimate of the overall total of Africans imported into the Americas would actually be somewhat lower than that of Curtin.[10] In spite of his drastic downward revision of

[9] Curtin, *Atlantic Slave Trade*, xv–xix, 272. [10] *Ibid.*, 86–87.

imports, one of Curtin's more startling conclusions is that between 1492 and 1770 more Africans arrived in the Americas than Europeans.

More meaningful than the all-embracing estimates of the gross total are the figures on how the imported slaves were distributed in various parts of the New World. The great bulk of them, nearly 90 percent, went to tropical America — to Brazil, to the Caribbean coast, and to the islands, the Greater and Lesser Antilles. Here lay the heart and center of the slave trade down through the centuries. The Caribbean Islands took some 4,040,000 of the total and South America some 4,700,000. When 175,000 are added to the total for the Old World traffic and 225,000 for Mexico, Central America, and Belize, 9,139,000 or 95.5 percent of gross estimated total of 9,566,000 for the entire Atlantic slave trade are accounted for.[11]

It is clear from this that British and French North America and subsequently the United States lay outside and beyond the thriving heart of the slave trade. Contrary to traditional accounts and schoolbook suggestions, as Curtin says, "the United States was only a marginal recipient of slaves from Africa." Allowing an estimate of 399,000 slaves imported by the British continental colonies and the United States from the beginning to 1861, and 28,000 for slaves taken into French (and Spanish) Louisiana, the number of slaves imported by the territory of the United States would come altogether to 427,000, or only 4.5 percent of the total. Of the several reductions in traditional estimates for the imports of the United States, the most drastic cutback that Curtin has made is in the previous estimates for illegal imports between 1808 and 1861, which he reduces to 54,000. Firm figures on these illegal imports admittedly require more study, but the revised estimate is persuasive. American slave ships were certainly active in the trade during these years, as well as others, but they appear to have landed their cargoes mainly in Cuba or in other foreign ports.[12]

[11] *Ibid.*, 88–89.

[12] *Ibid.*, 74–75, 88–89. Curtin thinks 54,000 for the years 1808–1861

One way to indicate the disparity between slave imports of the United States and those of other parts of Plantation America is to look at ratios between imports and geographical area. The islands of the Greater Antilles, for example, with an area roughly one third that of Texas, imported nearly six times the slaves landed in the entire territory of the United States. Of these islands, Saint Domingue, later named Haiti, which ended both slavery and slave trade by 1794, had by that date imported 864,000, or more than twice the total number of the United States. Jamaica had taken in 748,000 by the end of the legal trade in 1808, and Cuba, which continued to import Africans for another half century or more, received 702,000. In fact, Cuba took in more after 1808 than the United States received in all. The ratio of slave immigration to geographic area was even higher in the Lesser Antilles, and it should be remembered that the mountainous portion of most of the Antilles, great and small, is large. The Leeward Islands, a string of tiny tropical gems with a combined land area about half that of an average-size Mississippi county, imported 346,000 slaves — more than one thousand to the square mile. The highest ratio, more than twice that of the Leewards, was achieved by Barbados. Its 166 square miles served as the first (and for most, the last) New World home of 387,000 Africans. Two other heavy importers were the small French islands of Martinique, with 366,000, and Guadaloupe, with 291,000. These two islands together with Barbados and the Leewards comprise a land area less than twice that of Hinds County, Mississippi. Yet together they accounted for the importation of 1,390,000 African souls, considerably more than three times the number attributed to the continental expanse of British North America.[13]

The only American slave power readily comparable with the United States in geographic area is Brazil. With virtually half the continent of South America within its borders, Brazil is

is a "not unreasonable" estimate; see also Howard, *American Slavers and the Federal Law,* 255–257. [13] Curtin, *Atlantic Slave Trade,* 88–89.

larger in area than the continental United States minus Alaska, and only a little smaller with that state included. In comparing the two countries for this purpose, however, it should be remembered that while the "frontier" as officially defined had been closed in the North American republic about the time slavery ended in Brazil, the open and almost wholly unsettled lands of the latter country at that time still comprised much the greater part of the total land area. While the slave economy of the United States was concentrated in the Southern states, that of Brazil was more widely dispersed in the settled part of the country, but occupied no greater land area. It should also be kept in mind that African importations began nearly a century earlier and continued nearly a half-century longer in Brazil than they did in the United States. Estimates of Brazilian slave imports have been subject to wilder exaggeration than those of almost any other country, but have been drastically reduced by scholars in the last three decades. The most acceptable figure at present appears to be 3,647,000, or about eight and one-half times the number imported by the United States.[14]

Another way to indicate the special character of the African experience in the Southern states is to compare the rates of slave population increase there with rates of increase or decrease elsewhere. The long-term growth rate of the Afro-American population in the United States stands out as unique among the African diaspora in all nations of the New World. The rapid rate of increase was established early in the eighteenth century and continued throughout the period of enslavement. The decennial increase in the number of slaves after the close of the legal slave trade ranged from a low of 23.4 percent to a high of 30.6 percent between 1810 and 1860. The slave population alone, without the 10 to 13 percent of free Negroes, exceeded one million by 1810, was well over two million by 1830 and was very nearly four million in 1860, in addition to nearly a half million

[14] *Ibid.*, 47–49, gives a sketch of the historiography of Brazilian estimates; Curtin accepts the revisionist estimates without, in this case, attempting to go behind them.

free blacks in the latter census. By the time of emancipation, therefore, the 427,000 imported slaves had grown to a population more than ten times the number imported.[15]

Episodes of the great emancipation drama in other slave states spanned the century from 1794 to 1889. It opened in Saint Domingue, the most prized of France's overseas possessions, after the great slave revolt of 1791 (though slavery in a modified form was reestablished after independence). Saint Domingue had been the largest importer of Africans in the Caribbean. By the end of the slave trade and of slavery, she had taken in some 864,000 Africans, but the total slave population at that time was only 480,000. The abolition of slavery in the British West Indies in 1834 left a population of 781,000 freedmen or "apprentices," considerably less than half the aggregate of 1,665,000 slaves imported over the centuries.[16] By the time the British abolished slavery in the islands, their former colonies on the continent held about 2,200,000 in bondage. Had the slave increase in the British West Indies been as great in proportion to imports as it was up to that time in the continental slave states, the freed population would have been more than ten times the number on the slave rolls in 1834. Slave trade between the islands and with the continental countries, of course, complicates comparisons. The great majority of Southern slaves were imported directly from Africa. Among the islands some of the more striking contrasts are those between the 82,000 that Barbados liberated and the 387,000 she imported, between the 51,-300 the Leeward Islands held in 1834 and their 346,000 imports, and Jamaica's 311,000 freedmen compared with the 748,000 she had taken in.[17] The French islands of Martinique and Guadaloupe together imported a total of 656,600 slaves, but at the time of emancipation in 1848 they could not between

[15] U.S. Bureau of the Census, *Negro Population, 1790–1915* (Washington, D.C., 1918), 53.

[16] Curtin, *Atlantic Slave Trade,* 71.

[17] *Ibid.*, 59. On the origins of the South's slave imports, see U.S. Bureau of Census, *Historical Statistics of the United States . . .* (Washington, D.C., 1960), Series Z, 281–297, p. 769.

them account for one third that number of slaves to liberate. The Dutch possessions, with a bad reputation for health conditions and slave treatment, especially in Surinam, could count a surviving slave population at the end of slavery probably no greater than 20 percent of the roughly estimated half million they had imported.

Spanish and Portuguese colonies with the largest black populations, notably Cuba and Brazil, present peculiar problems that discourage similar comparisons between numbers imported and numbers liberated. In the first place, unlike the colonies of North Europeans, Cuba and Brazil went through phases of piecemeal emancipation during long wars (such as the South experienced in telescoped form during the Civil War) prior to abolition. By the time slavery was finally abolished, few slaves remained to be emancipated in either country, and the free black population greatly outnumbered the newly freed. The only meaningful way of testing and comparing the rates of growth among their Afro-American populations is through the use of racial statistics. This method is admittedly quite hazardous. Not only do the statistics vary in reliability and completeness, but there are wide variations in the social definition of the Negro.[18] In all countries the Afro-American population was swollen by miscegenation, but the increase from that source was affected not only by the extent of miscegenation, which varied greatly, but by the various ways in which the resulting genotypes were classified. The practice in the United States is clearly indicated by official instructions to enumerators in 1930: "A person of mixed white and Negro blood should be returned as a Negro, no matter how small the percentage of Negro blood." [19] But even if

[18] "The statistics of race are, then, in a rather chaotic state, and totals arrived at by indiscriminatingly adding the official figures, for example, of 'Europeans' in African colonies are most misleading." Kuczynski, *Population Movements* (cited in note 1), 86. See also the same author's *Colonial Population* (London, 1937), especially the chapter "Population by Race," 10–23.

[19] U.S. Bureau of the Census, *Census of the United States* (Washington, D.C., 1930); *Population*, II, 1398–1399.

the popular assumption is correct that miscegenation was more prevalent in Brazil and the percentage of people of mixed blood in the population was greater, the census in that country probably more than compensated for the difference by a less inclusive definition of the Negro and a greater tolerance for racial mobility and "passing." [20] In comparing populations of recent years one should also keep in mind the conclusion of an authority on the subject that "after the cessation of the slave trade the proportion of Negro population has declined steadily in most countries" of Latin America.[21]

Taking account of these hazards, conceding the extremely imprecise character of the data, but attempting to allow for them and arrive at the best estimates, Curtin ventures a suggestive comparison of the populations of African descent in various parts of the New World around the middle of the twentieth century.[22] The United States at that time, together with the small black population of Canada, accounted for some 31.1 percent of the New World population of African descent — with a start of only 4.5 percent of the total slave imports. The Caribbean Islands, on the other hand, starting with 43 percent of the total slave imports, contained only 20 percent of the Afro-American population at mid-century. Cuba took in 7.3 percent of the African immigrants, but harbored only 3 percent of the mid-century population. Brazil, the largest single importer of all, with a 38.1 percent share of the total, contained only 36.6 percent of twentieth-century Afro-Americans. Had the Brazilians of African descent increased in the same ratio to original imports that prevailed in the United States, they would have numbered 127,645,000 instead of 17,529,000 at mid-century, or more than two and a

[20] Wilbur Zelinsky, "The Historical Geography of the Negro Population of Latin America," *Journal of Negro History,* XXXIV (April 1949), 172–173, 188–189. [21] *Ibid.,* 207.

[22] The statistics Curtin mainly relies on are those of Angel Rosenblatt, an adaptation of which is most readily available in Frank Tannenbaum, *Slave and Citizen: The Negro in the Americas* (New York, 1946), in unnumbered pages between pages 15 and 16. Curtin's comparative table will be found in *Atlantic Slave Trade,* 91.

half times the estimated Afro-American population of the Western hemisphere. At the same ratio Afro-Americans in Cuba would have numbered 24,570,000 instead of 1,224,000. Applying this ratio of imports to increase in the Caribbean Islands as a whole would have resulted in an Afro-American population of 335,790,000 instead of the estimated 9,594,000 that inhabited the Antilles at mid-century.[23]

Some of these extrapolations, of course, appear completely fanciful in any real world of human ecology. Applying the same ratio between slave imports and present population in Barbados, for example, would people that green and lovely island with 13,545,000 expiring Afro-Americans — 80,000 of them per square mile — in place of the 1,000 per square mile that now eke out a living making rum and serving tourists in what is already, except for Malta, the most overpopulated island in the world.[24]

The point of all this is not an exercise in demographic fantasy, but simply one means of emphasizing a neglected historic experience — the uniquely rapid growth rate of the Afro-American population in the United States, more particularly in the

[23] In 1930 the foreign-born Negro population of the United States was less than one percent of the total Negro population, and the relative recency of this immigration was indicated by the fact that only 7.7 percent of those who survived in 1930 had arrived before 1900. In the census of 1870 there were only 9,494 foreign-born Negroes in the States, an undetermined number of whom were Africans imported by smugglers, and 28 percent of them were from Canada, whose population is included in the slave population of "British North America." Ira DeA. Reid, *The Negro Immigrant: His Background, Characteristics and Social Adjustment, 1899–1937* (New York, 1939), 42–43. In comparing post-slave-trade Negro populations of the United States and the Antilles, one should take into account the large number of Africans imported by the Caribbean islands after 1808, and in some islands much later, as "contract labor." These greatly outnumbered the illegally imported slaves and the Negro immigrants taken in by the United States. Wilhelmena Kloosterboer, *Involuntary Servitude Since the Abolition of Slavery: A Survey of Compulsory Labour Throughout the World* (Leiden, Holland, 1960), 191–203, and *passim.*

[24] David Lowenthal, "The Population of Barbados," *Social and Economic Studies,* VI, no. 4 (1957), 447.

slave states of that nation. The absence of early census data in sub-Saharan Africa precludes any comparison with growth rates in the mother continent, but no clue from what is known suggests population increases at all comparable. Certainly nothing even remotely comparable occurred among African diaspora in other nations of the world where data are available. Yet this unique American experience has gone virtually unnoticed or has been taken for granted by historians of the United States, even by those who specialize in comparative history of slavery and race relations. Frank Tannenbaum, for example, compares the way the Negroes prospered with the way the Indians "withered away and disappeared" in the West Indies; he compares the Negro in Cuba and Brazil with the Indian in Peru, and he makes many comparisons of the Negro's cultural advantages in Latin America with his disadvantages in Anglo-America.[25] But he neglects to make comparisons between the growth rate of the Negro population in the United States and the rates in other parts of the Americas. Gilberto Freyre also draws many illuminating comparisons between slavery and Negro life in Brazil and the Afro-American experience in the South, but he too, overlooks the glaring contrast in population increase between his native land and the northern republic.[26]

One likely reason for lack of special attention to the phenomenal rate of increase among Afro-Americans in the South is that it took place against the background of an unparalleled rate of increase among the white population of the United States as a whole, which has itself been neglected by investigation until recently. The enormous explosion in European populations and white populations around the world that began in the eighteenth century has long been appreciated and studied by demographers. Scholars have vigorously debated the causes of increase in Eng-

[25] Tannenbaum, *Slave and Citizen*, 40–42.

[26] Gilberto Freyre, *The Masters and the Slaves: A Study in the Development of Brazilian Civilization*, second ed. (New York, 1956); *The Mansions and the Shanties: The Making of Modern Brazil* (New York, 1963); *New World in the Tropics: The Culture of Modern Brazil* (New York, 1963).

lish population during the eighteenth and nineteenth centuries. Yet population growth in the United States of that period appears to have been about double the rate considered so remarkable in England. In his famous *Essay on the Principle of Population* (1798), Thomas Robert Malthus remarked on the enormous rate of American population growth, pronouncing it "a rapidity of increase, probably without parallel in history." In fact, it was specifically from the American phenomenon that he deduced the demographic law that bears his name.[27]

The Malthusian assumption that American population doubled every quarter of a century required an average decennial increase of 32 percent. The investigations of J. Potter, an English historian, indicate that the Malthusian rate was indeed roughly maintained through the eighteenth century and the first seven decades of the National period, and that it declined rapidly after 1860. Population increase was obviously swollen by two means other than natural increase, slave imports and white immigrants. Discounting increase from these sources, Potter arrives at the conclusion that the average rate of increase for whites was "slightly over 28 percent per decade" during the eighteenth century to 1790. During the next three decades data on immigration are too uncertain to make allowance for that source of increase. With allowance for the large white immigration between 1820 and 1860, the rate of natural increase for all, including blacks, in that period was 28.7 per decade.[28] The rate of decennial increase in the slave population of the United States after the close of legal imports from 1810 to 1860 was an average of 27.3 percent.[29] That rate almost equaled the national average, yet it was maintained in a region that embraced the two

[27] J. Potter, "The Growth of Population in America, 1700–1860," in D. V. Glass and D. E. C. Eversley, *Population in History* (London, 1965), 631–632.

[28] *Ibid.*, 643, 646, 666, 668. See also Yasukichi Yasuba, *Birth Rates of the White Population in the United States, 1800–1860* (Baltimore, 1961).

[29] U.S. Bureau of the Census, *Negro Population, 1790–1915,* 53. The rate of natural increase among free Negroes averaged in the same period only 21.5 percent per decade.

areas of highest mortality rates — the lower Mississippi Valley and the coastal plains of the Atlantic — as well as the four states with the highest death rates.[20] In spite of these handicaps and whatever handicaps may be attributed to their bondage, Southern slaves were increasing at a rate much greater than the population of any nation of Europe, nearly twice as fast as the population of England, and almost as rapidly as the white Americans in a period alleged by Malthus to have been "probably without parallel in history" for rapidity of increase. So far as history reveals, no other slave society, whether of antiquity or modern times, has so much as sustained, much less greatly multiplied, its slave population by relying on natural increase.

While historians have remained largely oblivious of these facts and the questions raised by them for more than a century, neither the facts nor their implications went entirely unmarked during the American crisis over the slavery issue. What is most surprising is that the defenders of Southern slave society did not make more use for their purposes of the information on comparative slave increase that was available to them at the time. James D. B. DeBow, the famous New Orleans editor and one time Superintendent of the United States Census Bureau, pointed out in passing in the *Compendium of the Census for 1850* a few of the more startling disparities between the increase of slaves in the South and their decrease in the West Indies, and later remarked that they had multiplied "more than five times faster than the population of France." [31] David Christy, a Northern journalist, was not so much a defender of slavery as he was an opponent of abolition and a champion of colonization, but he was often quoted and reprinted by pro-slavery propagandists. Christy did not exploit the population figures in his most popular work, *Cotton Is King,* published in 1855, but he did make some

[30] Potter, "Growth of Population in America," 678–679. Again, this assumes the revision of illegal imports, while admitting the need for further study.

[31] "The Vital Statistics of Negroes in the United States," unsigned article in *De Bow's Review,* XXI (1856), 405–410; *Compendium of the United States Census, 1850* (Washington, D.C., 1854), 84.

use of them two years later in his *Ethiopia,* a book promoting the cause of colonization in Liberia. He extended the comparison to Cuba and Brazil, as well as the West Indies, and speculated briefly on the causes of the difference, but made little of the matter.[32]

It was not a propagandist for slavery but a strong antislavery writer who went most deeply into the comparative figures on the slave trade and slave population and, somewhat to his discomfort, spelled out some of the questions raised by them. This was Robert Dale Owen, chairman of the American Freedmen's Inquiry Commission appointed in 1863 by Lincoln's Secretary of War Edwin M. Stanton to investigate the conditions and needs of the freedmen within the Union lines. It was the report prepared by this Commission that laid the foundation for the Freedmen's Bureau. Elaborating on the report, Owen published a book in 1864 entitled *The Wrong of Slavery, the Right of Emancipation* in which he explored such comparative statistics as he could find on slave imports and populations in the Americas. These figures, some of them quite close to and some still far from modern findings, posed a question that, he admitted, puzzled and disturbed him sorely:

> The answer involves results so extraordinary, at first sight so incredible — and, in effect, even when thoroughly examined, so difficult of satisfactory explanation — that I have devoted much time and labor to the critical revision of the materials whence my conclusions are drawn, before venturing to place them on record.
>
> This is the answer. *The half-million shipped for North America have increased nearly ninefold* — being represented in 1860 by a population exceeding four millions four hundred thousand; while *the fifteen millions sent to the West Indian colonies and to Southern America have diminished, from age to age,* until they are represented now by *less than half their original numbers!*
>
> How marvellous, beyond all human preconception, are these re-

[32] David Christy, *Ethiopia: Her Gloom and Glory as Illustrated in the History of the Slave Trade and Slavery . . .* (Cincinnati, 1857), 45, 91–92, 251.

sults! Had the fifteen millions whose lot was cast in the southern portions of our hemisphere increased in the same proportion as the half-million who were carried to its northern continent, their descendants, instead of dwindling to half, would have been today a multitude numbering more than a hundred and thirty millions of men!

Owen turns immediately to the inevitable question: "What is the explanation of this startling marvel?" The first hypothesis that occurs to him, that it could be due to "greater humanity with which the negroes of the United States have been treated, as compared with those of other slave countries," he initially rejects out of hand, first because of the evidence of ill treatment in the South and second for lack of solid information about treatment abroad. He then turns to alternative hypotheses such as might occur to any intelligent inquirer: the earlier end of the slave trade, the consequent differences in sex ratio and slave value and care, absentee ownership and overseer neglect of female care in other countries, the variables of climate and hygiene. He is too honest and intelligent to accept any of these hypotheses as adequate or very helpful explanations. Then, while taking care to point out that "as a general rule, wherever slavery exists at all," it is "essentially and degradingly evil," Owen abruptly returns to the hypothesis he initially rejected. After some reference to appalling rates of slave mortality in the West Indies, he writes: "Here [in the United States] that system had not borne its deadliest fruits. Here, especially for four or five decades after the Revolutionary War, certain features of a patriarchal character tended to alleviate the harshness." Coming from the pen of Robert Dale Owen in 1864, that is a remarkable statement. It might well have been lifted from the pages of his pro-slavery contemporary, George Fitzhugh of Virginia. In a footnote Owen cites testimony supporting his statement collected by the Freedmen's Inquiry Commission "in taking the evidence of freedmen, especially in the more northern Slave States." He does not elaborate on this theory, and he certainly does not offer the patriarchal theory as an adequate explanation. After worrying several hypotheses, he rejects them all as partial or inadequate and in effect abandons the effort to find a satisfactory explanation.[33]

The problem is, of course, still unsolved. The road to agreement on the correct explanation of the phenomenal rate of slave population increase will probably be longer and more painful than the road to agreement on the facts themselves. Scarcely a start has been made. No explanation will be attempted here. When one considers some of the difficulties and obstacles to agreement on this subject, the reasons for caution should be fully apparent. Few subjects could provoke more controversy, offer more temptations for invidious comparison, or arouse more defensiveness in racial, national and regional pride. For example, there will undoubtedly be those who are ready to seize upon these facts as prima facie evidence if not final proof of the superior mildness and humaneness of Southern slavery (as even Owen was tempted to do), the greater submissiveness and lack of rebellion among Southern slaves, and the superior benevolence and paternalism of Southern masters. At the other extreme will be those who are ready to point to the remarkable rate of increase among the slave population as nothing more than proof of the success of Southern slaveholders in the brutal and inhuman business of breeding slaves for the market.

Beyond these obvious obstacles to agreement and more difficult to solve are the problems of locating and weighing the almost unlimited number of influences that played a part in determining population growth. The mention of a few such problems will serve to suggest the complexities and difficulties involved. Without regard to priority or relative importance, there come to mind the equations of man-land ratios and man-woman ratios, land fertility and human fertility, rates of mortality and natality, the incidence of disease and famine, the relative advance of

[33] Robert Dale Owen, *The Wrong of Slavery, the Right of Emancipation and the Future of the African Race in the United States* (Philadelphia, 1864), 94–111. This book was called to my attention by Stephen Schreiber. See also George M. Weston, *The Progress of Slavery in the United States* (Washington, D.C., 1857), 83–91.

medicine and hygiene, comparative health conditions of tropical and temperate zones, tribal origins and mating customs of slave populations, and religious and legal traditions of masters. Complicating all these influences would be comparative stages of economic and technological development, transportation and food supply, fluctuations of world markets, and stages of boom and depression in local economies. On top of these would be added the imponderables of master-slave relations, the patriarchal ethic in tradition and practice: harshness of exploitation, severity of punishments, practices of cohabitation and miscegenation, discrimination between house servants and field hands, conditions of labor, and standards of feeding, clothing, and housing. Beyond these determinants lies the whole range of cultural limits to security of family, acceptance in the religious community, access to legal protection, and the possible avenues to economic independence, freedom, citizenship, and equality.

Problems of such scope and complexity are too numerous and difficult to be addressed here. Most of them still await the beginnings of investigation on a comparative scale. Perhaps the most useful contribution within the present range of possibilities would be some cautionary observations on the dangers of simplistic solutions.

The temptation to look for the answer in the comparative treatment of slaves is strong because it plunges immediately into the liveliest of historical controversies and the center of popular interest in the subject. Here we are assisted by a thoughtful essay of Eugene D. Genovese on the pitfalls of this subject for comparative historians. He points out that both Ulrich B. Phillips and Gilberto Freyre could be right in their contentions that slaves of their respective countries were the best treated, for they were thinking about quite different things. In speaking of "treatment" of slaves in the South, Phillips had in mind such material and measurable things as food, clothing, housing, and conditions of labor. On the other hand, in speaking of Brazil, Freyre referred predominantly to two other categories of "treatment": that having to do with opportunities for independence in social, economic, religious, and cultural life, and that having to do with access to freedom and citizenship.[34]

Which of these categories of treatment is most significant in explaining so complicated an equation as human reproduction and rates of population increase is debatable. Material considerations might seem uppermost if for nothing more than the elementary necessity to survive before one can reproduce. Statistics on European populations show an enormous decrease in childbearing in times of famine, but on the other hand rapid population increase in some of the countries with the lowest standard of living and poorest diets at the present time is a problem of grave concern. Rapid increase of population is neither an unmixed blessing, nor is it always correlated with material plenty. Good treatment of slaves in one sense, moreover, might imply bad treatment in another. Southern masters as a rule (though not always) supplied their slaves with food, while masters of the West Indies and elsewhere usually required their slaves to raise their own food after work hours and on weekends. But while the Southern slave benefited materially from this arrangement, he was at the same time deprived of the sense of property and the modicum of experience with entrepreneurship and independence gained by the produce-growing and market-going West Indian slave. The Southern slave was also deprived of many of the cultural, religious, and social opportunities as well as such chances of freedom and citizenship as were held out to Latin American slaves. But on the other hand an appalling number of Latin American slaves did not survive to enjoy these opportunities. If the free population of color was much greater and suffered fewer racial discriminations in Brazil than free Negroes did in the South, one consequence was that Brazilian planters of the nineteenth century often had to lock up their slaves tightly every night to keep them from running away and easily losing them-

[34] Eugene D. Genovese, "The Treatment of Slaves in Different Countries: Problems in the Applications of the Comparative Method," in Laura Foner and Eugene D. Genovese (eds.), *Slavery in the New World: A Reader in Comparative History* (Englewood Cliffs, N.J., 1969), 202–10.

selves among the free blacks. The consequence of better treatment for the free was therefore worse treatment for the enslaved blacks—inferior living quarters and less freedom.[35]

Since the areas of greatest mortality and lowest rate of increase lay in the tropics, and since the Southern slave states with their rapid rate of population growth were in the temperate zone, it is natural to look to climate for an explanation. It is clear that the tropics had special health problems and that tropical medicine developed too late to be of much help with them.[36] But before seizing upon tropical climate and disease as the seat of the trouble, one must take into account the experience of Colombia, Panama, and Ecuador. These countries took in only about 2 percent of the total slave imports, and yet they contained about 7 percent of the Afro-American population in 1950. The proportions were just about the reverse in Cuba, which accounted for over 7 percent of the total imports and only 3 percent of the total population of African descent at midcentury. Yet Cuba is closer to the temperate zone and Colombia, Panama, and Ecuador are much closer to the equator. Curtin has noted as another peculiarity of these three latter countries that 80 percent of their Afro-American population was mulatto, much higher than the percentage in other countries of the American tropics. In Haiti, for example, mulattos constituted only 10 percent of the Afro-Americans. Whether these racial differences are of any significance for the problem at hand remains to be investigated.[37]

Two variables of significance for this and many other aspects of comparative slave history deserve special attention. One of them is whether the local economy was booming or languishing; the other is whether the slave trade was open or closed. Both

[35] *Ibid.;* Kuczynski, *Population Movements* (cited in note 1), 35–37.

[36] Orlando Patterson, *The Sociology of Slavery: An Analysis of the Origins, Development and Structure of Negro Slave Society in Jamaica* (London, 1967), 98–103.

[37] Curtin, *Atlantic Slave Trade* (cited in note 2), 91, 93. Colombia and Panama, however, received considerable black migration after West Indian emancipation.

variables had a profound effect upon the exploitation and welfare of the slaves. In periods of boom and expansion in commercial agriculture, slaves were usually subject to greater exploitation and abuse than in periods of hard times and sluggish markets. Similarly in periods of open slave trade bondsmen were normally more abused and neglected than they were when the trade was closed and the supply of foreign replacements was cut off or diminished. Slave-labor economies, even those within the same empire or the same nation, differed greatly in the duration and intensity of boom periods and slave trade. Within Brazil a booming coffee frontier flourished while an old sugar economy declined, just as in the Southern slave states the old tobacco and rice plantations languished while a cotton and sugar frontier boomed.

The interval between the closing of the slave trade and the end of slavery also differed widely from place to place. In Haiti, there was no interval at all, since formal slavery and the slave trade ended almost simultaneously. In the British West Indies, the interval was twenty-six years or one generation, in other North European colonies somewhat shorter; in Cuba it was considerably less than a generation, and in Brazil somewhat more, though in both the latter countries the number of slaves and the stability of slavery declined along with the more gradual decline of the slave trade. It was in the slave states of the South that slavery flourished longest after legal imports were ended (longer, of course, in Virginia and Maryland than in the Lower South) — some fifty-seven years after the federal law prohibiting the trade and longer after the state laws against importation, in any case more than two generations. Instead of declining in number in the interval, as they did in Cuba and in Brazil and numerous other countries, the slaves of the Southern states quadrupled their numbers during the two generations from the end of the trade to the end of slavery. Whatever influence, great or small, good or bad, the shutting off of the sources of replacement may have had upon the character of the institution and the policies of the planters, that influence operated longer and upon larger populations in the South than it did anywhere else.

The temptation to attribute the unique rate of increase among the South's slaves to a deliberate policy of commercial breeding on the plantations will be irresistible for some. It is perfectly clear that planters stood to gain by the natural increase of their slave force, either for labor or for sale in the domestic slave trade. The early closing of African sources, the inadequacy of the original imports in view of the demand, and the rising price of slaves were added incentives. There is plenty of evidence of planters' interest in the rate of increase, and slave advertisements speak frankly of "good breeders." Abolitionists sometimes framed the charge of slave breeding in its most degrading form.[38] This was denied and heatedly resented in the South. Yet abolitionists were able to quote antislavery Southerners in support of their charge in some of its ugliest forms. A favorite quotation was from Thomas Jefferson Randolph, an opponent of slavery, who said in the Virginia debates of 1832 that slaves were "reared for the market, like oxen for the shambles." This barnyard imagery and conception of slave breeding has often been repeated, most recently in these words:

> Breeding consists of herding male and female livestock together in order to get an abundance of young pigs, lambs, colts, calves, or what you have for sale. That is exactly what slave owners did with the slaves, and there were no exceptions.[39]

[38] For example, see Theodore Dwight Weld, *American Slavery As It Is: Testimony of a Thousand Witnesses* (New York, 1839), 15, 182; John Elliott Cairnes, *The Slave Power* (London, 1862), 127–128, 134–135; Frederick Law Olmsted, *A Journey in the Seaboard Slave States* (New York, 1856), 157–158.

[39] Dwight Lowell Dumond, *Antislavery: The Crusade for Freedom in America* (Ann Arbor, 1961), 68. Dumond also quotes Thomas Jefferson Randolph, as did Weld in 1839, though like Weld and many since then Dumond attributed the quotation to Thomas Mann Randolph (who died in 1828), the father of Thomas Jefferson Randolph, the participant in the Virginia debates of 1832.

Slaveholders probably had less cause for resentment of the degrading implications of the livestock conception of human relations than did the slaves and their descendants, who surely had every right to be offended. But those planters with any experience at all would have known that more was involved than the crude elements of animal husbandry — the production of pigs, lambs, colts, and calves. For whatever reasons — and they were undoubtedly complex and numerous — slaves in the South reproduced themselves at a great rate, and those elsewhere did not. Some insight on the reasons might be gained from investigations of low rates of slave reproduction elsewhere.

The problems of reproduction rates and mortality rates are inseparable. As one historian of Jamaica put the matter, "the failure of reproduction as a means of assuring population growth among the slaves was to a large extent rooted in the prevailing levels of mortality." [40] In eighteenth-century Jamaica "too much labour, cruel punishments, starvation and general malnutrition were greatly responsible for the high mortality rates among slaves," and in particular, "heavy labour of the women was no doubt responsible for the frequency of gynaecological complaints among them." [41] Early in the century, planters abandoned hope of maintaining, much less supplying necessary growth in their slave population through natural increase:

> Vital processes could not ensure the population increase required for the plantations; indeed, without continued large-scale introduction of slaves, swift depopulation would have resulted. Probably any policy not explicitly based on the slave trade was foredoomed to failure. . . . It was economically impossible to maintain slavery as a profitable institution unless the rapidly wasting population was constantly recruited by importation of slaves. This could easily be done at the low prices at which they could be secured. Sir James Stewart considered the cost of rearing slaves in the West Indies was greater than the cost of constant importation of Africans. As

[40] George W. Roberts, *The Population of Jamaica* (Cambridge, 1957), 223. [41] Patterson, *Sociology of Slavery*, 103, 109.

H. M'Neill put it [in 1788], slaves were considered "as so many cattle" who would for seven or eight years "continue to perform their dreadful tasks and then expire." In short, slaves were expendable.[42]

Planters of Barbados imported three thousand Negroes a year in the eighteenth century, "but they worked their slaves so hard and fed them so skimpily that they died off as fast as new ones were brought in." Between 1712 and 1762 deaths exceeded births in the island by about 120,000, which meant an annual natural *decrease* of about 4.3 percent. Barbados has been called "the land of paradox . . . the richest and yet in human terms the least successful colony in English America." [43]

Attitudes toward reproduction, attitudes of both the planters and the slave women, had their consequences in birth rates and conditions that promoted or retarded childbearing and child rearing. Jamaican planters in the eighteenth century "strongly discouraged the breeding of children," and the "neglect, unhygienic conditions, and ill treatment of pregnant women led to many miscarriages and an excessively high death rate among children between 0 and 4 years of age." [44] It is the conclusion of an authority on population in the West Indies that:

> Low fertility was an inevitable result of the policy of relying on the slave trade and of the general conditions of slavery. Indeed, to the masters and slaves alike high-fertility patterns were unacceptable. The function of the female slave as a "work unit" was heavily stressed; in this capacity she was as essential to the plantation as the male slave. . . . It was even claimed by Governor Parry of Barbados, "the labour of the females . . . in the works of the field is the same as that of men." The rearing of children impaired her function as a labourer and thus was not countenanced by the

[42] Roberts, *Population of Jamaica,* 222.

[43] Richard S. Dunn, "The Barbados Census of 1680: A Profile of the Richest Colony in English America," *William and Mary Quarterly,* 3rd ser., vol. XXVI (January 1969), 25–26, 30; Lowenthal (cited in note 24), "Population of Barbados," 452.

[44] Patterson, *Sociology of Slavery,* 105, 111.

master. The position of the pregnant slaves, it seems, was not a happy one. In the words of Ramsay, they were "wretches who are upbraided, cursed and ill-treated . . . for being found in the condition to become mothers." A witness before the Select Committee of 1790–91 declared that "a female slave is punished for being found pregnant." [45]

For Brazil, the evidence of planter indifference or hostility to slave reproduction is, if anything, stronger than in the West Indies. One conditioning influence on Brazilian attitudes must have been the foreign slave trade, which kept the source of new slaves open nearly a half century longer than in the West Indies. It was cheaper and less trouble to import slaves than to raise them. Brazilian planters took no pains to balance the sexes among slaves and imported three or four times as many males as females. At some plantations no females at all appeared on the slave rolls.[46] On many others in the nineteenth century, the masters deliberately restrained slave reproduction by confining the sexes separately during the night.[47] Although much has been written about the solicitude of church and state for slave marriages in Brazil as compared with Protestant countries, Carl N. Degler concludes that "it is not likely that marriages of slaves in Brazil were any more enduring or protected from disruption through sale than in the United States." He points out that prior to 1869 there was no legal protection for the slave family in Brazil, and cites evidence that in any event an extremely small percent of the slaves were married or widowed.[48] Frightful rates of infant mortality and an excess of slave deaths over births are further evidence of indifference toward natural increase of the slave population. It is the conclusion of Degler, who believes that slavery in the United States was generally milder than in

[45] Roberts, *Population of Jamaica*, 225–26.

[46] Carl N. Degler, "Slavery in Brazil and the United States: An Essay in Comparative History," *American Historical Review*, LXXV (April 1970), 1018–1019.

[47] T. Lynn Smith, *Brazil: People and Institutions* (Baton Rouge, 1963), 130.

[48] Degler, "Slavery in Brazil and the United States," 1008–1009.

Brazil, that "Brazilians, in short, simply did not take sufficient care of their slaves for them to reproduce." [49]

A sampling of the attitudes and policies of Southern masters toward slave reproduction and child rearing might be of interest by way of comparison. The three examples of instructions to plantation managers taken from the mid-eighteenth century, the early nineteenth century, and the mid-nineteenth century cannot, of course, be assumed to be typical. Neither would they appear to be unrepresentative, and the frankly professed motives have little to do with benevolence. The first is from the instructions of James Semple, master of several Virginia tobacco plantations in 1759:

> The Breeding wenches more particularly you must Instruct the Overseers to be Kind and Indulgent to, and not force them when with Child upon any service or hardship that will be injurious to them & that they have every necessary when in that condition that is needful for them, & the children to be well looked after and to give them every Spring & Fall the Jerusalem Oak seed for a week together & that none of them suffer in time of sickness for want of proper care.[50]

The second sample is from the instructions of Thomas Jefferson, owner of more than 200 slaves, to his manager in 1819:

> . . . the loss of 5. little ones in 4 years induces me to fear that the overseers do not permit the women to devote as much time as is necessary to the care of their children: that they view their labor as the 1st object and the raising of their child but as secondary. I consider the labor of a breeding woman as no object, and that a child raised every 2. years is of more profit than the crop of the best laboring man. In this, as in all other cases, providence has made our interests and our duties coincide perfectly. . . . I must pray you to inculcate upon the overseers that it is not their labor, but their increase which is the first consideration with us.[51]

[49] *Ibid.*, 1020.

[50] Quoted in William Kauffman Scarborough, *The Overseer: Plantation Management in the Old South* (Baton Rouge, 1966), 183–184.

[51] Thomas Jefferson to Yancey, January 17, 1819, quoted in William

The third example is from the plantation rules of Andrew Flynn, a Mississippi cotton planter of the Yazoo Delta in 1840:

The children must be very particularly attended to, for rearing them is not only a Duty, but also the most profitable part of plantation business. . . . Pregnant women & sucklers must be treated with great tenderness, worked near home & lightly. Pregnant women should not plow or lift; but must be kept at moderate work until the last hour if possible. Sucklers must be allowed time to suckle their children from twice to three times a day according to their ages. At twelve months old children must be weaned.[52]

Both Brazilian and Jamaican planters took steps to encourage reproduction after the slave trade was closed, and Brazil looked to the Southern slave states for guidance. In spite of exhortations to follow the example of Virginians, however, the traditional arguments against raising slaves were still being advanced twenty years after the end of the African slave trade. A Brazilian planter reasoned in 1872 that:

One buys a Negro for 300 milries, who harvests in the course of the year 100 arobas of coffee, which produces a net profit at least equal to the cost of the slave; thereafter everything is profit. It is not worth the trouble to raise children. . . . Furthermore, the pregnant Negroes and those nursing are not available to use the hoe; heavy fatigue prevents the regular development of the fetus in some; in others the diminution of the flow of milk, and in almost all, sloppiness in the treatment of the children occurs, from which sickness and death of the children result. So why raise them?[53]

Jamaican slaveholders adopted "a new policy toward reproduction" toward the end of the eighteenth century that included "improving the conditions of pregnant women and of birth and

Cohen, "Thomas Jefferson and the Problem of Slavery," *Journal of American History,* LVI (December 1969), 518. According to this writer, who cites Jefferson's "Farm Book," "Between 1810 and 1822, about 100 slaves were born to Jefferson's 'breeding women'; while only a total of thirty Negroes died, were sold, or ran away." *Ibid.,* 519.

[52] Quoted in Scarborough, *The Overseer,* 69–70.

[53] Quoted in Degler, "Slavery in Brazil and the United States," 1018.

generally paying more attention to the raising of children." But in most cases "the unfavorable practices were too entrenched to be eradicated." [54] So were the attitudes of slave women, attitudes implanted by generations of abuse. In Jamaica there was evidence that "a considerable number of slave women disliked the idea of having children." Their attitude reflected those of the planters, who held out no promise of relief from field work during pregnancy or care for mothers and infants. Numerous observers "commented on the frequency of abortions." Marriages were rare in Jamaica and sexual abandonment the rule. The scarcity of white women in the island made black women more a prey to white men than usual. "The net result of all this," according to Orlando Patterson, "was the complete demoralization of the Negro male" who "eventually came to lose all pretensions to masculine pride and to develop the irresponsible parental and sexual attitudes that are to be found even today." [55]

If these evil effects of bondage upon mating and reproduction and parenthood were felt by slaves in the Southern states, and some of them doubtless were, they certainly did not manifest themselves in birth rates and population growth rates. If this were at least in some measure the result of "breeding" — of things done and not done by Southern planters — then the opposite effect in other slave societies was evidently to some degree the consequence of "nonbreeding." If the most significant variable were the duration of the slave trade and the availability of African replacements, it does not explain why the phenomenal rate of natural increase of slaves in the Southern states was established well before the end of the slave trade. If the policies of planters are responsible for the rate of population growth among Southern slaves, it remains to be explained how the natural increase of freed men kept up with that of whites after slavery. The rate fell for both, but the blacks generally kept up fairly well with the whites. If roughly equal sex-ratios among Southern

[54] Patterson, *Sociology of Slavery*, 106; Roberts, *Population of Jamaica*, 238–247.

[55] Patterson, *Sociology of Slavery*, 167–168, also 41–42, 159–166.

blacks explain much of the difference between their rate of natural increase and that of other Afro-Americans, it remains to be explained why sex-ratios elsewhere were so different — for they were not naturally determined. If sugar culture needed more male and less female labor than cotton culture, it is not clear from that why equal sex-ratios were established in the states before cotton culture was born and long before the slave trade ended. And finally if Malthusian law helps explain the remarkable rate of white increase in America, it might be worth asking whether it has anything to do with the rate of black increase.

These and many more questions need to be asked and answered before we find an acceptable explanation for the population growth of Afro-Americans in the United States. If the problem can be explored with sufficient detachment, patience, and skepticism, we may have here a more promising key to the comparative history of slavery and race relations than some that have already been used.

4

A Southern War Against Capitalism

If social theories regularly shared the fate of the social systems in which they were born, the history of thought would be a thin and impoverished thing of purely contemporary dimensions. The theories of George Fitzhugh came very near suffering the fate that befell the social order and institutions he defended. It was not merely that he was the spokesman of a cause that was overwhelmed in military disaster and an order that was leveled by revolutionary action. Nor was it simply that he was the outspoken champion of the discredited, despised, and abolished institution of Negro slavery. More important than the fall of the old order in explaining the eclipse of Fitzhugh was the sensational, if temporary, triumph of the system he opposed, a triumph that followed hard upon the collapse of the order he championed.

The fact was that the very aspects of "free society" that Fitzhugh most fiercely attacked, the aspects he repeatedly prophesied would spell the doom and downfall of that system, were the features that flourished most exuberantly in the decades following Appomattox. These features were an economy of laissez faire capitalism, an ethic of social Darwinism, and a rationalistic individualism of a highly competitive and atomized sort. Even part of his beloved South was later to join in the pursuit of these heresies.

It is little wonder that one writer could ask, "But who in

America would be reading Fitzhugh in twenty years?" The question was intended to be rhetorical and the answer was, of course, "Nobody." In the America of the post-Civil War period, admittedly, it is impossible to imagine a more completely irrelevant and thoroughly neglected thinker than George Fitzhugh.

The lapse of a century, however, has altered the perspective from which earlier generations assessed the significance of Fitzhugh's thought. The triumph of a highly individualistic society no longer seems as permanent in this country as it once did; nor does the disappearance of all forms of slavery before the advance of progress seem inevitable in the rest of the world. The current of history has changed again. Millions of the world's population are seeking security, abandoning freedom, and finding masters. It is not the sort of socialism that Fitzhugh advocated, nor the slavery he defended, but another type of system that he feared which is fulfilling his prophesies. Even in those societies where socialism is abhorred, mass production, mass organization, and mass culture render his insights more meaningful than they ever were in the old order of individualism.

It was Fitzhugh's constant complaint that his contemporary opponents rejected his theory out of hand without evaluation or understanding. He would have been more crushed by the total neglect of posterity, even in the South, until quite recently. For an intellectual tradition that stands in desperate need of contrast and suffers from uniformity — albeit virtuous liberal uniformity — this oversight is unfortunate. Granting his wicked excesses and sly European importations, Fitzhugh could at least furnish *contrast.* The distance between Fitzhugh and Jefferson renders the conventional polarities between Jefferson and Hamilton, Jackson and Clay, or Hoover and Roosevelt — all liberals under the skin — insignificant indeed. When compared with Fitzhugh, even John Taylor of Caroline, John Randolph of Roanoke, and John C. Calhoun blend inconspicuously into the great American consensus, since they were all apostles in some degree of John Locke.

With such a wealth of sterling and illustrious examples of the

Lockean liberal consensus, from Benjamin Franklin to Abraham Lincoln and on down, surely a small niche could be found in our national Pantheon for one minor worthy who deviated all down the line. For Fitzhugh frankly preferred Sir Robert Filmer and most of his works to John Locke and all his. He saw retrogression in what others hailed as progress, embraced moral pessimism in place of optimism, trusted intuition in preference to reason, always preferred inequality to equality, aristocracy to democracy, and almost anything — including slavery and socialism — to laissez faire capitalism. Whatever his shortcomings, George Fitzhugh could never, never be accused of advocating the middle way. Granting all his doctrine to be quite un-American, one might still ask that Fitzhugh's thought be reexamined, if only for the sharp relief in which it throws the habitual lineaments of the American mind.

Louis Hartz, who applauds America's rejection of Fitzhugh, has deplored the prevailing indifference to what he calls "The Reactionary Enlightenment" of the Southern conservatives. "For this was the great imaginative moment in American political thought," he writes, "the moment when America almost got out of itself, as it were, and looked with some objectivity on the liberal formula it has known since birth." While in his opinion the movement ran to fantasy, extravagance, and false identifications, he calls it "one of the great and creative episodes in the history of American thought," and its protagonists "the only Western conservatives America has ever had." [1]

Hartz is quite justified in placing Fitzhugh near the center and in the forefront of the Reactionary Enlightenment. He goes further to pronounce him "a ruthless and iconoclastic reasoner," "the most logical reactionary in the South," and to attribute to him "a touch of the Hobbesian lucidity of mind." He is on more doubtful ground when he pronounces the Virginian a "more im-

[1] Louis Hartz, *The Liberal Tradition in America* (New York, 1955), 147, 176. For an astute critique of Hartz and of Fitzhugh himself, published after the above was written, see Eugene D. Genovese, *The World the Slaveholders Made: Two Essays in Interpretation* (New York, 1969).

pressive thinker" than the Carolinian John C. Calhoun, but he qualifies his praise with numerous charges of inconsistency, irresponsibility, and even insincerity. In commenting upon the South's shift from the liberal doctrine of the Revolution to antebellum conservatism, Hartz writes: "Fitzhugh substituted for the social blindness of Jefferson a hopeless exaggeration of the truth. The South exchanged a superficial thinker for a mad genius." [2] I would not agree fully with either the praise or the indictment implied, but would cordially endorse the demand for serious attention to a neglected and provocative thinker.

It would be misleading, however, to leave the impression that George Fitzhugh was wholly typical of the Southern thinkers of his period or entirely representative of the proslavery thought or of agrarian thought. Fitzhugh was not typical of anything. Fitzhugh was an individual — *sui generis.* There is scarcely a tag or a generalization or a cliché normally associated with the Old South that would fit him without qualification. Fitzhugh's dissent usually arose out of his devotion to logic rather than out of sheer love of the perverse, but evidence warrants a suspicion that he took a mischievous delight in his perversity and his ability to shock. He once wrote teasingly to his friend George Frederick Holmes, referring to his *Sociology for the South,* "It sells the better because it is odd, eccentric, extravagant, and disorderly." [3] He was always a great one for kicking over the traces, denying the obvious, and taking a stand on his own.

For one thing, Fitzhugh was decidedly *not* an agrarian, for in his opinion "the wit of man can devise no means so effectual to impoverish a country as exclusive agriculture." Manufacturing and commerce were the road to wealth. "Farming is the recreation of great men, the proper pursuit of dull men." [4] As for that

[2] *Ibid.,* 159, 182, 184.

[3] Fitzhugh to Holmes, April 11, 1855, quoted in Harvey Wish, *George Fitzhugh, Propagandist of the Old South* (Baton Rouge, 1943), 126–127.

[4] George Fitzhugh, *Sociology for the South, or The Failure of Free Society* (Richmond, 1854), 15, 156.

sacred Southern dogma of free trade, it was a snare and a delusion, another fraud perpetrated by the Manchester heresy, to be avoided at all costs.[5] Those who dismiss Fitzhugh and his friends as bemused romantics enamored of feudalism will have to reckon with the Virginian's praise of Cervantes, who "ridded the world of the useless rubbish of the Middle Ages by the ridicule so successfully attached to it." [6] And those who might expect of him conventional Southern ideas of race will have to admit that he once deplored "the hatred of race" and anything that "cuts off the negro from human brotherhood," "because it is at war with scripture, which teaches that the whole human race descended from a common parentage; and, secondly, because it encourages and encites brutal masters to treat negroes, not as weak, ignorant and dependent brethren, but as wicked beasts, without the pale of humanity." [7]

Another quirk in Fitzhugh's philosophy that placed him outside the claim of typicality or representativeness — even of the class for which he spoke — was the very stress on the aristocratic and seigneurial prerogative and responsibility that was so central to his creed. What he so often failed to take into account and what those who picture him as authentic spokesman for Southern slave society fail to see is the intraracial egalitarianism (for all its shams) that was implicit in the "Greek Democracy." It was a white man's club, racially defined, and when aristocrats carried their pretensions much beyond excluding the black man (or carried their paternalism too far toward *in*cluding him) they were in trouble with the common whites — and most of them knew it very well. Mary Chesnut, after "one of Uncle Hamilton's splendid dinners, plate, glass, china, and everything," describes a revealing picture on the piazza where the gentlemen had retired

[5] *Ibid.*, 7–33.

[6] George Fitzhugh, *Cannibals All!, or Slaves Without Masters,* 132. All page references to *Cannibals All!* are to the John Harvard Library edition, C. Vann Woodward, ed. (Cambridge, Mass., 1960).

[7] Fitzhugh, *Sociology,* 95, 147; but see also 262, 264, 265, 271, indicating contrary tendencies. He had all along believed in the innate inferiority of the Negro, and later became a Negrophobe.

for their cigars. In their midst sat Squire Macdonald, the well-digger: "clay pipe in his mouth, he was cooler than the rest, being in his shirt sleeves, and he leaned back luxuriously in his chair tilted on its two hind legs, with his naked feet up on the bannister." And worst of all, "See how solemnly polite and attentive Mr. Chesnut is to him!" [8]

Apart from his ideas, Fitzhugh had traits of personality and character that discourage classifying him with any type. In many ways he was the antithesis of the fierce-eyed, grim-faced polemicist who stares out from the picture galleries of the 1850's, whether in the wing for Southern fire-eaters or the wing for abolitionists. Fanaticism is not compatible with a temperament that selects Falstaff and Sancho Panza as favorite characters of fiction. He made much of his remote family connection with the prominent abolitionist leaders Gerrit Smith and James G. Birney, and his acquaintance with other abolitionists. He sought them out, cultivated them. "We have an inveterate and perverse penchant of finding out good qualities in bad fellows," he wrote. "Robespierre and Milton's Satan are our particular friends." [9] There was none of the suspicious recluse in him. "We admire them all, and have had kindly intercourse and correspondence with some of them," he said of the abolitionists. He referred often to his debate with Wendell Phillips and to the "generous reception and treatment we received, especially from leading abolitionists, when we went north to personate Satan by defending Slavery." [10] Even after the war, when his world was in ruins, his home part of a battlefield, and his enemies were plotting more mischief, he could write in the old vein: "Love is a pleasanter passion than hate, and we have been hating so intensely for the last six years, that we are now looking about for something to love. . . . We are resolved to hate no one, and to quarrel with no one. No, not even with Thad. Stevens and his men." [11] If

[8] Mary Boykin Chesnut, *A Diary from Dixie* (ed. Ben Ames Williams, Boston, 1961), 143.

[9] Fitzhugh, *Cannibals All!*, 97. [10] *Ibid.*, 86.

[11] George Fitzhugh, "Thad. Stevens's Conscience — The Rump Parliament," *De Bow's Review*, After the War Series, II (1866), 469–470.

candor and magnanimity could disarm hostile critics, Fitzhugh was well endowed.

George Fitzhugh was born in Prince William County, Virginia, on the Northern Neck between the Potomac and the Rappahannock rivers on November 4, 1806. He sprang from a numerous family that included men of large landed property and prominence in the history of Virginia. He was descended from William Fitzhugh, "a fair classical scholar, a learned, able, and industrious lawyer, a high tory, high Churchman," who came to the colony in 1671 as land agent for Lord Fairfax. George Fitzhugh's father, a doctor and small planter, did not prosper, and the paternal estate passed out of family hands shortly after his death in 1829. That year, however, the son improved his lot somewhat by marrying Mary Brockenbrough of Port Royal, Caroline County, and promptly moving into his wife's home. This was described later by an unsympathetic neighbor as a "rickety old mansion, situated on the fag-end of a once noble estate." [12]

Like the mansion, the village of Port Royal had seen better days, but it was prettily seated on the banks of the Rappahannock, and Fitzhugh became devoted to the village and its few citizens. Its most distinguished citizen, the great Jeffersonian intellectual, John Taylor of Caroline, had died five years before young Fitzhugh moved to Port Royal. It is not known whether Fitzhugh ever met Taylor, but he did have several of the famous agrarian's books in his library and their influence may be detected in his own works. He acquired several slaves through his marriage, practiced law in a desultory way, and fathered a family of nine children.[13] He was a confirmed homebody. "Love and veneration for the family is with us not only a principle, but probably a prejudice and a weakness. We were never two weeks at a time from under the family roof, until we had passed middle life, and now that our years almost number half a century, we have never been from home for an interval of two months." [14]

[12] Wish, *Fitzhugh,* 1–13. [13] *Ibid.,* 14–18.
[14] Fitzhugh, *Cannibals All!,* 192.

His one visit in the North came in 1855, when he visited Gerrit Smith in Peterboro, New York, and debated Wendell Phillips in New Haven.

In formal education, Fitzhugh never progressed beyond the old field school. His learning in the law he picked up on his own. His real education came of his independent, undirected, unsystematic, but wide reading. "We are no regular built scholar — have pursued no 'royal road to mathematics,' nor to anything else," he confessed. "We have by observation and desultory reading, picked up our information by the wayside, and endeavored to arrange, generalize, and digest it for ourselves." [15] Unlike many defenders of the South, he took pains to read the opposition. "We have whole files of infidel and abolition papers, like the *Tribune,* the *Liberator* and *Investigator,*" he reported. "Fanny Wright, the Devil's Pulpit and the Devil's Parson, Tom Paine, Owen, Voltaire, *et id genus omne,* are our daily companions." [16] He also read some of the British economists, as well as a few of the English and Continental socialists. In the main, however, he relied upon conservative British journals to keep him abreast of current thought and literature, periodicals such as the *Edinburgh Review,* the *Westminster Review, Blackwood's Magazine,* and the *North British Review.* More in keeping with his station and his time were his love of the Latin classics and his habit of quoting from them.

Of all his contemporaries, Thomas Carlyle seems to have made the most profound impression upon Fitzhugh. The Scot was then at the height of his popularity, and his Virginia admirer quoted him often and with relish, especially his diatribes against the cant of philanthropists and his Olympian thunder against the "Dismal Science," its professors, and their miserable "laws of the shop-till." Carlyle's essay, "The Present Age," furnished the text and suggested both title and subtitle of *Cannibals All!* [17] Carlyle's attack upon the British abolitionists and West Indian

[15] *Ibid.,* 67. [16] *Ibid.,* 192.

[17] In Thomas Carlyle, *Latter-Day Pamphlets* (London, 1850). See 52–53, 56.

emancipation was naturally grist for the pro-slavery propaganda mill, but Fitzhugh was interested in the broader implication of Carlylean doctrine, particularly his diatribes against the "Mammonism" of the industrialists and the wickedness of Manchester economics. The Virginian identified himself willingly with Young England, Disraeli, and Tory socialism.

Fitzhugh's activity in politics was only local, and he never stood for elective office. He was acquainted with a number of Virginia politicians, however, and managed to ingratiate himself with President Buchanan, who appointed him law clerk in the office of the Attorney-General. His literary apprenticeship, late in starting, was served first as a writer for the *Federalsburg Democratic Record,* 1849–1851, and later as editorial writer for the *Richmond Examiner* in 1854–1856 and for the *Richmond Enquirer* in 1855–1857. His most significant journalistic writing was for *De Bow's Review,* edited by the fiery champion of the South, James D. B. De Bow of New Orleans. According to the reckoning of Harvey Wish, Fitzhugh published "well over a hundred articles" in that journal between 1855 and 1867. He also placed occasional essays in the *Southern Literary Messenger* of Richmond, and after the war in *Lippincott's Magazine* of Philadelphia and the *Southern Magazine* of Baltimore.[18]

The political crises of Europe and America in 1848 and 1849 appear to have stimulated Fitzhugh's first independent publications, two pamphlets that appeared in Richmond in 1850 and 1851 and were later reprinted as an appendix to *Sociology for the South.* In the first and more important of them, *Slavery Justified,* he announced several themes that he was to repeat and elaborate in later works. "Liberty and equality," he declared in his opening sentence, "are new things under the sun." France and the Northern states of the union, the only parts of the world that had given the combination an extensive trial, had proved the experiment was "self-destructive and impracticable" and had "already failed." The evidence of failure was the social distress, economic suffering, and political revolt in both countries.

[18] Wish, *Fitzhugh,* 343–344.

"How can it be otherwise," he asked, "when all society is combined to oppress the poor and weak minded?" Since " 'Every man for himself, and devil take the hindmost' is the moral which liberty and free competition inculcate," and since "half of mankind are but grown-up children" it was apparent that "liberty is as fatal to them as it would be to children." What the weak needed was protection, that is, masters, and so they turned to socialism and communism. Plantation slavery of the South was "the beau ideal of Communism; it is a joint concern, in which the slave consumes more than the master . . . is far happier . . . is always sure of a support . . . and is as happy as a human being can be." To call free labor "wage slavery" as the socialists did was "a gross libel on slavery," for the condition of free labor was "worse than slavery." The wage system was a contradiction of human needs. "Wages are given in time of vigorous health and strength, and denied when most needed, when sickness or old age has overtaken us. The slave is never without a master to maintain him." The consequence was that "at the slaveholding South all is peace, quiet, plenty and contentment. We have no mobs, no trades unions, no strikes for higher wages, no armed resistance to the law, but little jealousy of the rich by the poor. We have but few in our jails, and fewer in our poor houses." [19]

The one cloud in this otherwise idyllic picture of Southern felicity was the free Negro, the few hundred thousand of their race who were deprived of the protection and security of slavery. Their miserable condition was proof of the curse that freedom would prove to their race: another experiment in liberty that ended in failure. The free Negroes of the North were "an intolerable nuisance," and resentment against them there had provoked stringent discriminatory laws and proscription. The answer to the question that formed the title of his second pamphlet, *What Shall Be Done with the Free Negroes?,* was, return

[19] Fitzhugh, *Slavery Justified — Liberty and Equality — Socialism — Young England — Domestic Slavery,* reprinted in Appendix of Fitzhugh, *Sociology for the South,* 226–258.

them to slavery. The Negro was fitted only for that status. "We have fully and fairly tried the experiment of freeing him . . . and it is now our right and our duty, to listen to the voice of wisdom and experience, and reconsign him to the only condition for which he is suited." [20]

For all his professed love of learning and devotion to logic, Fitzhugh was a propagandist and not infrequently displayed the tactics of that craft, as well as an occasional sample of its disingenuousness. He excused the irregularity of his warfare on the ground that it was necessitated by the nature of his adversaries. "They are divided into hundreds of little guerrilla bands of Isms," he said, "each having its peculiar partisan tactics, and we are compelled to vary our mode of attack from regular cannonade to bushfighting, to suit the occasion." And so he did, shifting his ground, masking his batteries, and resorting to a variety of tricks to confuse his opponents. He deliberately adopted a style, he said, "in which facts, and argument, and rhetoric, and wit, and sarcasm, succeed each other with rapid iteration." [21] More serious was the damaging admission he made in a personal letter to his friend Professor Holmes in 1855. "I assure you, Sir," he wrote, "I see great evils in slavery, but in a controversial work I ought not to admit them." [22]

Fitzhugh's first book, published in 1854, was also the first published in America bearing the new word "sociology" in its title.[23] *Sociology for the South, or The Failure of Free Society* opened with an aggressive assault upon Adam Smith, laissez faire, and all the political economists who advanced the proposition that social well-being was "best prompted by each man's eagerly pursuing his own selfish welfare unfettered and unrestricted by legal regulations, or governmental prohibitions. . . ." He pronounced that philosophy "false and rotten to the core."

[20] Fitzhugh, *What Shall Be Done with the Free Negroes?*, reprinted in *ibid.*, 259–306. [21] Fitzhugh, *Cannibals All!*, 239.

[22] Fitzhugh to Holmes, April 11, 1855, quoted in Wish, *Fitzhugh*, 111.

[23] Henry Hughes of Mississippi published *A Treatise on Sociology* (Philadelphia, 1854) the same year.

Such a system not only opened the way for the rich and strong to exploit the poor and weak individuals, but under the guise of "free trade" it paved the way to empire by enabling the industrial and commercial economies to exploit countries with agricultural economies. "Thus is Ireland robbed of her very life's blood, and thus do our Northern states rob the Southern." [24]

The remedy for these ills was not less government but more government. "Government may do too much for the people, or it may do too little," he thought. "We have committed the latter error." [25] Like his contemporary, Henry C. Carey of Philadelphia, Fitzhugh advocated a system of vigorous government participation in economic development, but he urged it upon Virginia and the Southern states as a solution to *their* problems. Both his economic and social aims and his means of attaining them were departures from the Jeffersonian tradition. He stressed the social values of manufacturing and commerce and the need for the growth of cities in the South to foster these arts. Government should be employed vigorously for planning and promoting schemes of internal improvements, developing financial and marketing facilities, and fostering transportation, particularly the building of railroads. "Our system of improvements, manufactures, the mechanic arts, the building up of our cities, commerce, and education should go hand in hand." Above all it was important to provide public education. "Poor [white] people can see things as well as rich people. We can't hide the facts from them. . . . The path of safety is the path of duty! Educate the people, no matter what it may cost!" [26]

After disposing of the classical economists to his satisfaction, Fitzhugh turned to the socialists. He was concerned here mainly with English and French socialists, and he treated them with a great deal more respect than he had Adam Smith and the classical economists. Few realized, he wrote "how much of truth, justice and good sense, there is in the notions of the Communists, as to the community of property." They fully acknowledged

[24] Fitzhugh, *Sociology*, 11–14; see also 133. [25] *Ibid.*, 145.
[26] *Ibid.*, 144–148; see also 153.

the obligations of society to the weak and the propertyless and exhibited a sense of responsibility and morality that was foreign to the capitalistic economists. Socialism was, after all, only "the new fashionable name for slavery." If classical economics was "the science of free society," he would pronounce socialism "the science of slavery." Slavery's only quarrel with socialism was the refusal of the latter to acknowledge fully the failure of free society and its futile attempt to build upon institutions and ideas of a discredited order. "We slaveholders say you must recur to domestic slavery," he insisted, "the best and most common form of Socialism. The new schools of Socialism promise something better, but admit, to obtain that something, they must first destroy and eradicate man's human nature." [27] In that the socialists shared the fallacies of the philosophers who founded free society.

The trouble started with John Locke and the Enlightenment of the eighteenth century. "The human mind became extremely presumptuous" in that era, he wrote, "and undertook to form governments on exact philosophical principles, just as men make clocks, watches or mills. They confounded the moral with the physical world, and this was not strange, because they had begun to doubt whether there was any other than a physical world." Under their spell, Jefferson and a few misguided patriots embellished an otherwise healthy colonial rebellion with the abstractions of the Declaration of Independence and the Virginia Bill of Rights. These principles were "wholly at war with slavery" and, "equally at war with all government, all subordination, all order," in fact. "Men's minds were heated and blinded when they were written, as well by patriotic zeal, as by a false philosophy, which, beginning with Locke, in a refined materialism, had ripened on the Continent into open infidelity." For a long time these abstractions were a dead letter, had little effect, and did little harm, since inherited English institutions continued with little change. "Those institutions were the growth and accretions of many ages," he pointed out, "not the work of legis-

[27] *Ibid.*, 42, 47, 61, 72.

lating philosophers." But now that the abolitionists were inflamed with these notions, Locke and Jefferson should be firmly refuted and repudiated.[28]

Fitzhugh opened up on the "self-evident truths" and "inalienable rights" with the zeal of an iconoclast. What was really self-evident was that "men are not born physically, morally, or intellectually equal," and that "their natural inequalities beget inequalities of rights." "It would be far nearer the truth to say, 'that some were born with saddles on their backs, and others booted and spurred to ride them' — and the riding does them good." The ideal of equality was not only false but immoral. "If all men had been created equal, all would have been competitors, rivals, and enemies." Nature had a better plan. "Subordination, difference of caste and classes, difference of sex, age, and slavery beget peace and good will." Life and liberty were obviously not "inalienable" since "they have been sold in all countries, and in all ages, and must be sold so long as human nature lasts." [29] His distant kinsman, George Mason, author of the Virginia Bill of Rights, had carried self-deception to the point of solemnly forbidding "such harmless baubles as titles of nobility and coats of arms," and this in the face of "the right secured by law to hold five hundred subjects, or negro slaves, and ten thousand acres of land, to the exclusion of everybody else . . . an exclusive hereditary privilege far transcending any held by the nobility of Europe. . . ." This was arrant hypocrisy. "We have the *things,* exclusive hereditary privileges and aristocracy, amongst us, in utmost intensity; let us not be frightened at the *names*. . . ." [30]

Fitzhugh did not place his faith in paper institutions — declarations, bills of rights, or even written constitutions — but in organic, flesh-and-blood institutions:

> State governments, and senators, and representatives, and militia, and cities, and churches, and colleges, and universities, and landed property, are institutions. Things of flesh and blood, that know their

[28] *Ibid.*, 175–177. [29] *Ibid.*, 177–183. [30] *Ibid.*, 184, 190–191.

rights, 'and knowing dare maintain them.' We should cherish them. They will give permanence to government, and security to State Rights. But the abstract doctrines of nullification and secession, the general principles laid down in the Declaration of Independence, the Bill of Rights, and Constitution of the United States, afford no protection of rights, no valid limitations of power, no security to State Rights. The power to construe them, is the power to nullify them.[31]

"Institutions are what men can see, feel, venerate and understand"; such institutions as were associated with Moses, Alfred, Numa, and Lycurgus. "These sages laid down no abstract propositions, founded their institutions on no general principles, had no written *constitutions.* They were wise from experience, adopted what history and experience had tested, and never trusted to *a priori* speculations, like a More, a Locke, a Jefferson, or an Abbé Sieyès." The teaching of history was "that it is much better, to look to the past, to trust to experience, to follow nature, than to be guided by the *ignis fatuus* of *a priori* speculations of closet philosophers." [32]

Encouraged by the reception of his *Sociology* in "the confidence that we address a public predisposed to approve our doctrine, however bold or novel," [33] Fitzhugh plunged into the most productive period of his life. Between 1854 and 1857 he was not only writing editorials for the *Richmond Examiner* and *Richmond Enquirer* and articles for *De Bow's Review,* but he was also at work on his major book, *Cannibals All! or, Slaves Without Masters,* published in Richmond in 1857.

If he considered the *Sociology* "bold" and "novel," he undoubtedly thought of its successor as still bolder and even more novel. The very title, *Cannibals All!* proclaimed the moral relativism with which he proposed to affront conventional American values. He realized at the outset, he said, that "our analysis of human nature and human pursuits is too dark and sombre to meet with ready acceptance." [34] Nothing daunted, he set out to

[31] *Ibid.,* 188–189. [32] *Ibid.,* 183, 187.
[33] Fitzhugh, *Cannibals All!,* 7. [34] *Ibid.,* 35.

perform a sort of Nietzschean transvaluation of American values, a subversion of the national faith in progress and the goodness of human nature, as well as the characteristic addiction to liberalism, optimism, and respectability. "But if you would cherish self-conceit, self-esteem, or self-appreciation," he warned, "throw down our book; for we will dispel illusions which have promoted your happiness, and show you that what you have considered and practiced as virtue is little better than moral Cannibalism." In the Virginian's inverted hierarchy of values, freedom was slavery and slavery freedom, respectability was criminal and crime respectable. His premise was that "all good and respectable people are 'Cannibals all,' who do not labor, or who are successfully trying to live without labor, on the unrequited labor of other people: — Whilst low, bad, and disreputable people, are those who labor to support themselves, and to support said respectable people besides." [35]

Fitzhugh was saying what Thorstein Veblen unconsciously paraphrased nearly half a century later: "Those employments which are to be classed as exploit are worthy, honorable, noble; other employments, which do not contain this element of exploit, and especially those which imply subservience or submission, are unworthy, debasing, ignoble." [36] Like Veblen, the Virginia sociologist accounted reputability proportional to guile: "The more scalps we can show, the more honored we are." But unlike him, Fitzhugh made no pretense of detachment and moral neutrality. "You are a Cannibal!" he charged, "and if a successful one, pride yourself on the number of your victims, quite as much as any Fiji chieftain, who breakfasts, dines and sups on human flesh." [37]

Fundamental to his critique of free society and his defense of slavery was an extreme form of the labor theory of value, which he had absorbed, he said, from the socialists. "My chief aim," he

[35] *Ibid.*, 16.

[36] Thorstein Veblen, *The Theory of the Leisure Class: An Economic Study in the Evolution of Institutions* (New York, 1899), 15.

[37] Fitzhugh, *Cannibals All!*, 39, 17.

wrote in his Preface, "has been to show, that *Labor makes values, and Wit exploitates and accumulates them;* and hence to deduce the conclusion that the unrestricted exploitation of so-called free society, is more oppressive to the laborer than domestic slavery." [38] His attack was directed at the moral complacency and assumption of superiority in the North. "We are, all, North and South, engaged in the White Slave Trade," he insisted, "and he who succeeds best, is esteemed most respectable. It is far more cruel than the Black Slave Trade, because it exacts more of its slaves, and neither protects nor governs them." Proof of it lay in the admittedly greater profitability of free labor, which to Fitzhugh only meant that the employer of free labor retained more and gave labor less of the value created than did the owner of slave labor. "You, with the command over labor which your capital gives you, are a slave owner — a master, without the obligations of a master. They who work for you, who create your income, are slaves, without the rights of slaves. Slaves without a master!" In his opinion the masters of free labor "live in ten times the luxury and show that Southern masters do," while "free laborers have not a thousandth part of the rights and liberties of negro slaves." On the other hand, "The negro slaves of the South are the happiest, and, in some sense, the freest people in the world." [39]

In Fitzhugh's philosophy the idea of progress was a modern delusion. Modern history, in fact, was a record not of progress but of retrogression. He advanced the theory in his *Sociology* that "the world has not improved in the last two thousand, probably four thousand years, in the science or practice of medicine, or agriculture," and that it had actually been "retrograding in all else save the physical sciences and the mechanic arts. . . . It is idle to talk of progress, when we look two thousand years back for models of perfection." [40] Bourgeois criteria were "purely utilitarian and material," typified by the sentiments and philosophy of Benjamin Franklin, which were "low, selfish, atheistic and

[38] *Ibid.*, 5. [39] *Ibid.*, 15, 17–19. [40] Fitzhugh, *Sociology*, 157–159.

material." Acceptance of these criteria tended "to make man a mere 'featherless biped'; well-fed, well-clothed and comfortable, but regardless of his soul as 'the beasts that perish.' "[41]

In *Cannibals All!* he repeated his attack on the idea of progress and elaborated his theory of retrogression in the arts, but to this he added a rejection of the Whiggish interpretation of history. He grounded his critique on something rather similar to the Marxian dialectic of class struggle. He interpreted each political upheaval in terms of material advantages or disadvantages accruing to conflicting social classes. The theory advanced in William Blackstone's *Commentaries* that the appearance of the House of Commons near the reign of Henry III "was the dawn of approaching liberty," he rejected out of hand. "We contend that it was the origin of the capitalist and moneyed interest government, destined finally to swallow up all other powers in the State, and to bring about the most selfish, exacting and unfeeling class despotism." The emancipation of the serfs was not "another advance toward equality of rights and conditions," as Blackstone claimed, because "it aggravated inequality of conditions, and divested the liberated class of every valuable, social, and political right." Blackstone was also wrong in holding that the Reformation "increased the liberties of the subject," for "in destroying the noblest charity fund in the world, the church lands, and abolishing a priesthood, the efficient and zealous friends of the poor, the Reformation tended to diminish the liberty of the mass of the people, and to impair their moral, social and physical well-being."[42]

In leveling his sights on the Glorious Revolution of 1688, Fitzhugh was striking close to the philosophical underpinnings of the American Revolution. His interpretation of the events of 1688 was similar to that later advanced by Marx. Far from "the consummation or perfection of British liberty" usually pictured, the Glorious Revolution in Fitzhugh's view was "a marked epoch in the steady decay of British liberty." What really happened was that the powers of the House of Commons were in-

[41] *Ibid.*, 90. [42] Fitzhugh, *Cannibals All!*, 107.

creased at the cost of prerogatives of the Crown, the church, and the nobility, "the natural friends, allies, and guardians of the laboring class." The settlement and the subsequent chartering of the Bank of England "united the landed and moneyed interests, placed all the powers of government in their hands, and deprived the great laboring class of every valuable right and liberty. The nobility, the church, the king were now powerless; and the mass of people, wholly unrepresented in the government, found themselves exposed to the grinding and pitiless despotism of their natural and hereditary enemies." The subsequent history of Britain was another story of "slaves without masters," the degradation of the masses. He quoted Charles Dickens as saying, "Beneath all this, is a *heaving* mass of poverty, ignorance and crime." Those who persisted in describing this sad decline as "progress" were afflicted with a willful blindness.[43]

It is of possible significance that upon the title pages of his two books, the subtitles, *Failure of Free Society* and *Slaves Without Masters,* are printed in larger type than the main titles. At any rate, the disparity suggests the predominance of attack over defense in the author's polemics. In both works, the great bulk of space is given over to the shortcomings and failing of free society. The ferocity of Fitzhugh's indictment of the capitalist economy surpasses that of John Taylor at the beginning of the century and the Southern Populists toward its close, and is equaled only by the severity of the socialist attack. He could snarl at "this vampire capitalist class" as bitterly as any socialist. In fact, Fitzhugh exploited and quoted extensively from many of the sources that Karl Marx used ten years later in the first volume of *Capital* to marshal evidence of the inhumanity of British industrial capitalism. Prominent among these sources were the reports of Parliamentary Commissions appointed to investigate conditions among coal, iron, and textile workers in the 1840's. He also relied upon Tory reformers and British journals, particularly the *Edinburgh Review,* which he called "a grand repository of the ignorance, the crime, and sufferings of the

[43] *Ibid.,* 108. Cf. Karl Marx, *Capital* (Chicago, 1919), I, 795–796.

workers in mines and factories . . . in fine, of the whole laboring class of England." [44]

It was perhaps the abundance of material at hand that caused Fitzhugh to concentrate mainly upon British rather than American conditions. He made no more allowance than did Marx for the occasional exaggerations of reformers and spared his readers none of the horrors of the evidence. One is treated to the full measure of misery in the Scottish coal pits with their subhuman child laborers from five to thirteen years of age, who did not see the light of day for weeks on end; their naked men and "almost naked" women workers harnessed to coal carts and toiling twelve to fourteen hours a day. There were the child calico-printers of Lancashire who were kept at their tasks for fourteen or sixteen consecutive hours and grew up with pipe-stemmed legs, pinched faces, and brutalized minds. There were the metal workers of Wolverhampton "cruelly beaten with a horsewhip, strap, stick, hammer handle, file, or whatever tool is nearest at hand." [45] Was it any wonder, asked Fitzhugh, that an American abolitionist after intensive study of British labor conditions declared that "he would sooner subject his child to Southern slavery, than have him to be a free laborer of England." [46] If the degraded workers of England were to be enslaved, in fact, they would "by becoming property, become valuable and valued" and would be elevated from their present plight at least to the status of domestic animals.[47]

In his attack upon free society in America, Fitzhugh borrowed freely from the tactics Northern abolitionists employed in their propaganda against Southern slavery. The criminology of the subject was employed as fair description of conditions, and the occasional instance of sadism and depravity presented as the prevailing practice. Just as the abolitionists made effective use of quotations from Southerners and slaveholders upon the evils of the slave society, so Fitzhugh employed to the fullest the testimony of abolitionist reformers upon the failing of free society.

[44] Fitzhugh, *Cannibals All!*, 158. [45] *Ibid.*, 167–176.
[46] *Ibid.*, 109. [47] *Ibid.*, 155.

He described the testimony of "the actual leaders and faithful exponents of abolition" whom he quoted as "our trump card." Out of their own mouths he would prove "the inadequacy and injustice of the whole social and governmental organization of the North." Since many of the abolitionists were passionate critics of free society and entertained schemes for reform of the North as well as the South, each with his own panacea, Fitzhugh did not lack for live ammunition from their publications.[48]

It was good propaganda to catalogue the desperate remedies and wild panaceas to which the breakdown of free society had driven the reformers, and Fitzhugh did so regularly: "Mr. Greeley's Phalansteries, Mr. Andrews' Free Love, Mr. Goodell's Millennium and Mr. [Gerrit] Smith's Agrarianism," to say nothing of the multitudinous nostrums of Mr. Garrison, "King of the Abolitionists, Great Anarch of the North." [49] Why all the remedies if there were no diseases in free society?

> Why have you Bloomer's and Women's Right's men, and strong-minded women, and Mormons, and antirenters, and "vote myself a farm" men, Millerites, and Spiritual Rappers, and Shakers, and Widow Wakemanites, and Agrarians, and Grahamites, and a thousand other superstitious and infidel Isms at the North? Why is there faith in nothing, speculation about everything? Why is this unsettled, half-demented state of the human mind co-extensive in time and space, with free society? Why is Western Europe now starving? and why has it been fighting and starving for seventy years? Why all this, except that free society is a failure? Slave society needs no defence till some other permanently practicable form of society has been discovered. Nobody at the North who reads my book will attempt to reply to it; for all the learned abolitionists had unconsciously discovered and proclaimed the failure of free society long before I did.[50]

It was not the slaveholders but the abolitionists who "have for years been roaring . . . to the Oi Polloi rats, that the old crazy

[48] *Ibid.*, 85, 93–95. [49] *Ibid.*, 211.
[50] Fitzhugh to A. Hogeboom, January 14, 1856, *ibid.*, 103.

edifice of society, in which they live, is no longer fit for human dwelling, and is imminently dangerous." [51] Yet the same philosophers were inviting the South to abandon its stable society, demolish its tested and benevolent institutions, and move into the rickety edifice of free society which the abolitionists had already condemned as uninhabitable.

Fitzhugh's books appeared at the height of the sectional controversy over slavery and the literary war it induced. It is all the more difficult for that reason to assess their relative importance and distinguish their influence from that of the many other proslavery books of the period. His *Sociology* profited much from the wholehearted endorsement and several notices it received from George Frederick Holmes, foremost reviewer in the South. De Bow and other Southern editors gave it space and frequent notice, and the first printing of the book was almost sold out in a few months. Fitzhugh complained of "the affectation of silent contempt" by the Northern press, and such attention as he got in that quarter was in the main hostile. Among Southern writers who revealed the influence of the *Sociology,* whether admitted or not, Professor Wish mentions Edmund Ruffin (who did acknowledge the impact of Fitzhugh's "novel and profound views"), Albert Taylor Bledsoe, William J. Grayson, James D. B. De Bow, George D. Armstrong, and Thornton Stringfellow.[52]

The response to *Cannibals All!* was also flattering in the South, though some of its critics in that quarter were troubled by its concession to socialism and by its carelessness and inconsistencies. Even De Bow thought the author "a little fond of paradoxes, a little inclined to run a theory into extremes, and a little impractical." But on the whole the Southern reception was enthusiastic. The book, after all, marked an advance to "higher ground" in the Southern position at the climax of the sectional controversy. It virtually ignored defensive tactics and concen-

[51] *Ibid.*, 96.

[52] Wish, *Fitzhugh,* 126; and for summary of critical reception of *Sociology* (cited in note 4) see 113–125.

trated upon aggressive strategy along radical anticapitalist lines.[53]

The legend of the satanic champion of evil from Port Royal had grown in antislavery circles since the appearance of his first book, and *Cannibals All!* was received as the last word in diabolism. William Lloyd Garrison, according to Professor Wish, gave it "considerably more attention than perhaps any other book in the history of the *Liberator*." Garrison quoted long passages from the book as horrible examples of the extremes to which "this cool audacious defender of the soul-crushing, blood-reeking system of slavery" could go as the spokesman of "the cradle-plunderers and slave-drivers at the South." Surely *Cannibals All!* could be written down as the "gospel according to Beelzebub that is preached at the South." In another mood he called Fitzhugh "the Don Quixote of Slavedom — only still more demented" than the Knight of La Mancha. "If he is not playing the part of a dissembler, he is certainly crack-brained, and deserves pity rather than ridicule or censure." But Garrison spared him neither censure nor ridicule: the Virginian was "demoniacal," his writings "idiocy." Granted that Fitzhugh showed a certain ingenuity and cleverness in pointing out the failings of free society, these were as nothing compared to the abominations of slave society. Granted the South might be free of "Isms" — so was despotic Russia.[54]

Many of Fitzhugh's more startling phrases and paradoxes, taken out of context, lent themselves admirably to quotation by Republicans or antislavery people for the purpose of discrediting the South or the Democrats. The more extreme pronouncements of the Virginian were thereby represented erroneously not only as typical of *his* views, but also those of his party and his region. Abraham Lincoln was a faithful reader of Fitzhugh's articles

[53] *Ibid.*, 198. The criticism of *Cannibals All!* is summarized in 195–199.

[54] *The Liberator,* March 6, 13, April 19, 1857, quoted in Wish, *Fitzhugh,* 200–203.

and editorials in the *Richmond Enquirer* and cited them in his speeches during the years 1856 to 1859 as representative of the wicked purposes of Democrats and slaveholders. He even went so far as to connive in planting one of the *Enquirer* articles in a proslavery paper of Springfield, which could then be quoted to the embarrassment of local Democrats.[55] William Herndon, his law partner, bought a copy of *Sociology for the South* and reported that "this book aroused the ire of Lincoln more than most proslavery books." If he read as far as page 94 in that work, Lincoln found this passage: "One set of ideas will govern and control after a while the civilized world. Slavery will every where be abolished, or every where be re-instituted." He could have found the same idea in different words in *Cannibals All!* (page 106). Actually, in replying to charges of Senator Stephen A. Douglas regarding the famous House Divided speech of 1858, Lincoln said in a speech at Cincinnati, September 17, 1859: "But neither I, nor Seward, nor [Congressman John] Hickman is entitled to the enviable or unenviable distinction of having first expressed that idea. The same idea was expressed by the Richmond 'Enquirer' in Virginia, in 1856, quite two years before it was expressed by the first of us." Lincoln mistakenly attributed the irrepressible conflict idea to Roger A. Pryor, editor of the paper, but the author of the unsigned editorial of May 6, 1856, to which he referred was really Fitzhugh.[56]

The strain of irresponsibility in his writings involves Fitzhugh in the guilt of his generation. It is somewhat ironical, however, that the Virginian should have been associated in such a personal way with the origins of the irrepressible conflict and House Divided ideas, or that he should have sometimes been made a symbol of Southern intransigence and militant disunionism. He actually deplored nullification, opposed secession till the last moment, and dreaded disunion. In all probability, he no more

[55] Albert J. Beveridge, *Abraham Lincoln, 1809–1858* (Boston, 1928), II, 30–31.

[56] See Wish, *Fitzhugh,* 150–159, 288, on the whole House Divided speech matter.

desired a bloody showdown than did Lincoln. The final words of *Cannibals All!* were a friendly appeal to the abolitionists:

> Extremes meet — and we and the leading Abolitionists differ but a hairbreadth. . . . Add a Virginia overseer to Mr. Greeley's Phalansteries, and Mr. Greeley and we would have little to quarrel about. . . . We want to be friends with them and with all the world; and, as the curtain is falling, we conclude with the valedictory and invocation of the Roman actor — "Vos valete! et plaudite!"

Like many Americans of his day, George Fitzhugh was absorbed in the great sectional crisis, and his mind was molded by the titanic surge and flow of the struggle. Unlike all but a very few, however, he managed to achieve a modicum of timelessness and universality in his theories that places them just beyond the destructive reach of historical relativism and should spare them dismissal as the cynical rationalization of outmoded institutions. In the opinion of Charles A. Beard, Fitzhugh's thought was "unlimited in its sweep," a "universal view" that "exceeded the planting view and the agrarian view in its scope of details, in its diversity of content, and in its reach of time." [57] He would seem to deserve some attention outside the context of slavery and the Civil War.

In so far as the phrase is permissible at all, Fitzhugh was an American original. For many reasons he rejected the political principles of the Enlightenment upon which Jefferson based his thought. He found the economic determinism of Madison inadequate for his purposes, as well as the narrow capitalist-versus-agrarian dialectic of the other sage of Port Royal, John Taylor. The parochialism of Calhoun appealed to him, but not the Carolinian's legalistic constitutionalism, and most certainly not his heretical adaptation of Lockean theory. Unlike his Southern predecessors and contemporaries, Fitzhugh's approach was not political, or economic, or legalistic, but sociological and psychological — both with an antique flavor, and yet more attuned to the modern than to the eighteenth- or nineteenth-century

[57] Charles A. Beard, *The American Spirit* (New York, 1942), 288.

mind. His belated discovery of his intellectual obligations to the ancients is revealing of both his naïveté and his originality. He wrote Professor Holmes in 1855:

> I received from Mr. Appleton's, a week ago, Aristotle's Politics and Economics. I find I have not only adopted his theories, his arguments, and his illustrations, but his very words. Society is a work of nature and grows. Men are social like bees; an isolated man is like a bird of prey. Men and society are coeval. . . . Now, I find that, although Locke, Rousseau, Adam Smith, Jefferson, Macaulay, and Calhoun are against me, Aristotle, Carlyle, you, and all the leading minds of the day are with me. . . . I used to think I was a little paradoxical. I now fear I am a mere retailer of truisms and commonplaces.[58]

If man were a social and political animal from the word go, and if society were an organism — as organic as an elaborated and diversified beehive — then all the talk about compacts, and social contracts, and man in a state of nature, and natural rights, and consent of the governed, and equality was arrant nonsense. Hobbes was as wrong as Locke, and Jefferson as wrong as Calhoun, and Adam Smith was out of his mind. Society was an organic continuum, inegalitarian and hierarchical. The inequalities explained and necessitated the hierarchy. They also required that government recognize the facts of life, refuse to be blinded by laissez faire dogmas, and intervene to protect the weak from the strong. An ethic of devil-take-the-hindmost led straight to tyranny and anarchy. Inequality was both necessary and desirable, and stability was important above all. This did not exclude change. Growth meant change, and society was a growing organism. But it had to grow according to the laws of growth, slowly, with uninterrupted continuity of institutions and moral values — not in fits and starts and revolutions according to the specifications and theories of philosophers and utopian dreamers. "Such is the theory of Aristotle," he wrote, "promulgated more than two thousand years ago, generally considered

[58] Fitzhugh to Holmes, April 11, 1855, quoted in Wish, *Fitzhugh,* 118–119. See also *Cannibals All!,* 12–13.

true for two thousand years, and destined, we hope, soon again to be accepted as the only true theory of government and society." [59]

There came a point, however, where Fitzhugh departed from the Aristotelian way. While man was social, he was not rational He was, in fact, fundamentally irrational, guided not by reason but by instinct, custom, habit, and requiring tradition and religion and stable institutions to keep him in line. Fitzhugh held and often expressed a profound skepticism of all atomistic and rationalistic theories of human nature and a strong aversion to rationalistic philosophers. "Modern philosophy treats of men only as separate monads or individuals," he complained.[60] His attitude was not that of anti-intellectualism, but he distrusted intellect that departed from experience and history and tradition and arrogantly spun lofty abstractions. Especially did he distrust intellect fired by moral passion and conviction of self-righteousness. "In fine, all of the greatest and darkest crimes of recorded history have been perpetrated by men 'terribly in earnest' blindly attempting to fulfill, what they considered, some moral, political or religious duty." There had been too much of that type of intellect loosed upon mankind, he wrote after the war, too much in the South as well as the North.[61]

In renouncing Locke and rationalism and the Enlightenment, Fitzhugh had no notion of renouncing the heritage of the American Revolution — the true heritage, that is. Not at all, he declared. "All the bombastic absurdity in the Declaration of Independence about the inalienable rights of man, had about as much to do with the occasion as would a sermon or oration on the teething of a child or the kittening of a cat. . . . Our institutions, State and Federal, imported from England where they had grown up naturally and imperceptibly . . . would have

[59] Fitzhugh, *Cannibals All!*, 71. See also Arnaud B. Leavelle and Thomas I. Cook, "George Fitzhugh and the Theory of American Conservatism," *Journal of Politics*, VII (1945), 145–168.

[60] Fitzhugh, *Cannibals All!*, 54.

[61] Fitzhugh, "Terribly in Earnest," *De Bow's Review*, After the War Series, II (1866), 172–177.

lasted for many ages, had not thoughtless, half-informed, speculative men, like Jefferson, succeeded in basing them on such inflammable materials. . . . The Revolution of '76 was, in its action, an exceedingly natural and conservative affair; it was only the false and unnecessary theories invoked to justify it that were radical, agrarian and anarchical." These were the theories of John Locke, "a presumptuous charlatan, who was as ignorant of the science or practice of government as any shoemaker or horse jockey." These theories had not inspired the Revolution of '76, but by slow fuse had eventually touched off "the grandest explosion the world ever witnessed," the Revolution of '61. "The French Revolutions of '89, 1830, and 1848, were mere popguns compared to it; as we all see and feel, for its stunning sound is still ringing in our ears." He wrote that in 1863 when, indeed, the stunning sound of revolution was deafening.[62]

While Fitzhugh expressed distrust for any book on moral science less than four hundred years old, he made one significant exception, Sir Robert Filmer's *Patriarcha.* This was the work of the forgotten Kentish monarchist and conservative whom Locke had felt it necessary to belabor extensively in his *Treatise.* Filmer's stress upon the patriarchal family, rather than his defense of the divine right of monarchs, was what struck the deep responsive chord in Fitzhugh. For the Virginian the family was everything, and society, government were but the family writ large — the authoritarian, patriarchal family. Aristotle had taught him "that the family, including husband, wife, children, and slaves, is the first and most natural development of that social nature." It was the model of all institutions: "this family association, this patriarchal government . . . gradually merges into larger associations of men under a common government or ruler." It was the disproof of Locke, for "fathers do not derive their authority, as heads of families, from the consent of wife

[62] Fitzhugh, "Revolutions of '76 and '61 Contrasted," *ibid.,* IV (1867), 36–42. This was originally published in the *Southern Literary Messenger,* XXXVII (1863), 718–726.

and children." It was the justification of domestic slavery, for "besides wife and children, brothers and sisters, dogs, horses, birds and flowers — slaves also belong to the family circle," and there they, like other weaker members, received the care, protection, and control they needed.[63]

There is a possible clue in Fitzhugh's thought to the provocative questions and indictments advanced by Louis Hartz in his treatment of the Reactionary Enlightenment. Because Hartz finds "beneath the feudal and reactionary surface of Southern thought" nothing but slavery, he concludes that the massive structure of reaction was "a simple fraud" and that, "Fraud, alas, was the inevitable fate of Southern social thought." He goes on to say, "They exchanged a fraudulent liberalism for an even more fraudulent feudalism: they stopped being imperfect Lockes and became grossly imperfect Maistres. This was the meaning of Fitzhugh's 'great conservative reaction'. . . ." Not only did their system not fit the American liberal formula, "the real trouble with it was that it did not fit any formula, any basic categories of Western social theory." [64]

Fitzhugh did find a formula, and he did not go back to feudalism nor forward to Maistre for it — only back to the seventeenth century, to John Locke's chosen antagonist, Sir Robert Filmer. And in seventeenth-century Kent, as well as in Virginia of that and the two following centuries, as Peter Laslett has pointed out, the patriarchal family was a pretty "basic category." That was why Locke took Filmer seriously. "Filmerism," says this historian, "was above all things the exaltation of the family: it made the rules of domestic society into principles of political science." [65] The gentry of Kent were a close-knit community. "The genealogical interrelationships between its members were extensive, complicated and meticulously observed by

[63] Fitzhugh, *Cannibals All!*, in order of quotation, 193, 72, 243, 205.
[64] Hartz, *The Liberal Tradition*, 146–148.
[65] Peter Laslett, "Sir Robert Filmer: The Man and the Whig Myth," *William and Mary Quarterly*, ser. 3, V (1948), 544.

all of them: it is astonishing how distant a connexion qualified for the title 'cozen.' The reason for this excessive consciousness of kinship was patriarchalism." [66]

This Kentish cousinage and its "excessive consciousness of kinship" extended across the Atlantic. "By 1660," writes Laslett, "this group of interrelationships existed in two places in the world: in middle eastern Kent and in Virginia in the area of the James River. This process, by which an English county society reproduced its names, its attitudes, its literary interests, even its field sports, in the swamps of the Virginia creeks, had begun with the foundation of the Virginia Company of London in 1606. . . . The story of the early Virginian planters' families illustrates the most important feature of the English gentry of the time — the immense strength of the family bond and the extraordinary cohesion of the grouping of families by locality. There could be no more vivid illustration of patriarchalism at work." [67] As for the Kentish gentry, "The most characteristic thing they produced was the political thinking of Sir Robert Filmer and the most surprising was the society of the Old South in the United States." [68] In fact, the transplanted patriarchal offshoot outlasted the original: "the descendents of the Virginian planters, who became the slaveholders of the Southern States, were the heads of a classic type of patriarchal household, so that it survived until the middle of the nineteenth century even in such a rationalistic and egalitarian society as the U.S.A." [69] The English historian even suggests that the Southern branch had "lineaments even more strongly marked than in England, perhaps because there were then no towns of any consequence." [70]

But what of the Revolutionary generation of the Virginia Enlightenment — the Masons, the Randolphs, the Jeffersons, the Madisons, the Washingtons — and their apparently firm com-

[66] Peter Laslett, "The Gentry of Kent in 1640," *Cambridge Historical Journal,* IX (1948), 150. [67] *Ibid.,* 161–162. [68] *Ibid.,* 162–163.

[69] Peter Laslett (ed.), Introduction to Sir Robert Filmer, *Patriarcha and Other Political Works* (Oxford, 1949), 26.

[70] Laslett, "The Gentry of Kent," 150n.

mitment to the antithesis of Filmerism, to the Lockean principles of rationalistic individualism and its picture of an atomistic society? Whatever principles these gentry subscribed to in the 1770's and later, they were one and all patriarchs on their own — Thomas Jefferson included. The anthropological, sociological, and political realities of Virginia society were those of the patriarchal family. Those realities might better be understood in Filmerian than in Lockean terms. And curiously enough, many of these Revolutionary Lockeans were blood relatives — remote "cozens" of Sir Robert himself. "In Virginia, the Filmers and the Filmer connections," writes Laslett, "were associated with all the great families which finally gave to the thirteen colonies their Revolutionary leadership in the 1770's — the Washingtons, the Byrds, the Berkeleys, and the Randolphs and so the Jeffersons. Whatever the subsequent literary and philosophical reputation of Sir Robert Filmer, he had been a great genealogical success." [71] Sir Robert was a family man in more ways than ideological.

It was not "simple fraud" that led George Fitzhugh to seize upon Filmer in his search for some ideological basis on which to construct his defense and his understanding of Virginia society, even in mid-nineteenth century. As a sociologist he had got hold of some firm anthropological data. It is rather more a wonder that the patriarchs of Revolutionary Virginia should have temporarily embraced Locke than that their sons should have returned to Filmer.

In an elaborate comparison, Fitzhugh identified the South and its heritage and tradition with Filmer, and the North and its tradition and heritage with Locke. The English Tories, with whom he identified the South, "are conservative, for the most part, agreeing with Sir Robert Filmer"; while the English Whigs "are progressive, rationalistic, radical, and agree with Locke in his absurd doctrines of human equality and the social contract." These were "the antinomes or opposing forces" in the mother

[71] Laslett (ed.), Introduction to Filmer, *Patriarcha,* 10.

country: Filmer *versus* Locke. "The North and the South would pretty well supply the places, or act the part, of these forces in America." [72]

Fitzhugh's home at Port Royal was shelled by Union troops during the war, while he and his family were refugees in Richmond. After the war and for more than a year during the Johnsonian Reconstruction, he was employed, oddly enough, as an agent of the Freedmen's Bureau and served with a Negro freedman as an associate judge of the Freedmen's Court. From this vantage point he viewed the Reconstruction drama and wrote about it philosophically, sometimes humorously, but rarely with any bitterness. After the death of his wife in 1877, he moved to Kentucky to live with a son, and finally to Texas with an impoverished daughter. There he died, nearly blind, in 1881 at the age of seventy-four.

In the many articles he published in *De Bow's Review* after the war, Fitzhugh counseled the South to accept the new order, but he showed little disposition to retract any of his antebellum theories. In 1867 he had the hardihood to reprint an article he had written in 1863 at the crest of the Confederate tide: "We begin a great conservative reaction," he had announced. "We attempt to roll back the reformation in its political phases, for we saw everywhere in Europe and the North reformation running to excess, a universal spirit of destructiveness, a profane attempt to pull down what God and nature had built up and to erect ephemeral Utopia in its place." [73]

In 1857 he had defiantly addressed to the abolitionists a boast of the security and confidence of the Old Regime: "Is our house tumbling about our heads, and we sitting in conscious security amidst the impending ruin?" he asked. "No! No! Our edifice is one that never did fall, and never will fall; for Nature's plastic

[72] Fitzhugh, "The Impending Fate of the Country," *De Bow's Review*, II (1866), 569.

[73] Fitzhugh, "Revolutions of '76 and '61 Contrasted," *ibid.*, IV (1867), 43.

hand reared it, supports it, and will forever sustain it." [74] So far as the record reveals, he never had the hardihood to reprint that, but as a philosopher he may have reflected upon it from time to time.

[74] Fitzhugh, *Cannibals All!*, 97.

5

The Northern Crusade Against Slavery

FROM the opposite ends of American history has come evidence of white racism, of both its antiquity and its persistence. Winthrop D. Jordan unearthed its origins in Elizabethan England and sixteenth-century Europe and traced its growth as the functional rationale of white supremacy and American identity down to 1812. The Kerner Report spelled out the disastrous consequences in the violence and riots in contemporary America.[1] But what of the period between? Was there not, as legend has it, an interlude of redeeming virtue in the mid-nineteenth century when white Americans, inspired by the antislavery crusade, put aside their racism, rededicated themselves to their ideals of equality, and waged a heroic war for freedom and a temporarily successful campaign for racial equality? Or was the crusade itself corrupted and frustrated by a sickness endemic among the crusaders?

Answers to these questions are obscured by time and propaganda, by vested interests of racial and national pride. For one thing, the justification of the bloodiest war in national history, a war resulting in the sacrifice of 600,000 lives, more than the number of Americans killed in two world wars, is at stake. Answers will be slow in coming and may never be perfectly clear.

[1] Winthrop D. Jordan, *White Over Black: American Attitudes Toward the Negro, 1550–1812* (Chapel Hill, N.C., 1968), and *Report of the National Commission on Civil Disorders* (Washington, D.C., 1968).

From time to time, however, additional insights are provided by historians, even when they have other purposes and problems in mind.

One illuminating source of insight is Eileen S. Kraditor's study of the abolitionist strategy and tactics.[2] To her surprise she came out with a new and favorable revision of the prevailing interpretation of William Lloyd Garrison. She began with the received opinions, presented most recently in two able biographies of Garrison, both published in 1963.[3] As she says, whatever respect they inspire for their subject is "more than balanced by the conviction that he was bullheaded, arrogant, vindictive, and incredibly blind to some obvious truths." For more than a generation it has been the practice, even among the most strongly pro-abolitionist historians, to protect the reputation of the movement by disavowing Garrison's importance or centrality in it. Dwight L. Dumond, for example, though an ardent champion of abolitionists, puts Garrison down as "insufferably arrogant" and (in italics) "a man of distinctly narrow limitations among the giants of the antislavery movement." [4]

Miss Kraditor does not contend that Garrison was a typical abolitionist or that he represented majority opinion on antislavery strategy. Nor does she deny his personal idiosyncracies and foibles (though she does put in a timid and not too convincing claim for his sense of humor). But she is "struck by the logical consistency of his thought on all subjects," granted his principles, with which she finds herself usually in agreement. She admits that he changed his opinions from time to time but holds that "the changes themselves represented a logical development." Though she does not use the terms, she sees Garrison as

[2] Eileen S. Kraditor, *Means and Ends in American Abolitionism: Garrison and His Critics on Strategy and Tactics, 1834–1850* (New York, 1969).

[3] John L. Thomas, *The Liberator: William Lloyd Garrison, A Biography* (Boston, 1963), and Walter M. Merrill, *Against the Grain: A Biography of William Lloyd Garrison* (Cambridge, Mass., 1963).

[4] Dwight L. Dumond, *Antislavery: The Crusade for Freedom in America* (Ann Arbor, 1961), 173–74.

the "hedgehog" (in Sir Isaiah Berlin's sense) among the "foxes," the man who "knows one big thing." His big thing was that abolitionism was a *radical* and not a *reform* movement. It is a bit anomalous to find a Marxian historian applying the term "radical" to a bourgeois thinker like Garrison, who never questioned the capitalist system, "free enterprise," or "free labor." What she means is that Garrison, unlike his opponents, believed that slavery and the racial dogmas which justified it so thoroughly permeated American society and government, North as well as South, that the eradication of the institution and its ideological defenses — and the racism of the latter was as important to him as slavery — was a root-and-branch operation. On that he never equivocated.

Garrison's abolitionist opponents were reformers, not real "radicals," even in the limited sense of the term. They believed in constitutional means and political strategy. They professed to be "realists" and sought to attract moderates rather than repel them by extremism and "extraneous" issues. They believed that American society, government, and institutions were fundamentally sound and that once the alien institution of slavery was removed, all would be well. Hence they were appalled at Garrison's intransigent denunciation of the Constitution as "a covenant with death, and an agreement with hell" which "should be immediately annulled." They deplored his demand for disunion, along with sundry "extraneous" demands, such as no government, no church, and no party. Moderates believed him capable of following to its conclusion "every corollary, and every corollary of every corollary of a syllogism." To them there were limits to logic.

To the incorrigible radical, logic has few limits. Slavery was a sin and that was that, and the only thing to do about sin was to stop sinning. Now! As for the impracticability of his demands, his answer was that politics was the art of the possible, and that his role was agitation, the art of the desirable. To ask the agitator to trim his demands for the sake of expediency was to miss the point. Garrison's ends were much too radical for political and parliamentary means. In his opinion American society was not fundamentally sound but thoroughly corrupt, top to bottom. Down with it, root and branch. He would not trim, he would not compromise, he would not vote for corrupt politicians or support corrupt governments and churches, and he would not temper his means to his ends.

The old Liberator will find more unqualified admirers on the contemporary scene than he would have a few years ago. But even the most hot-gospel root-and-brancher of today will have difficulties with the Garrisonian rhetoric and premises. As Miss Kraditor says, "The key to Garrison's ideology is perfectionism." He believed implicitly in the perfectability of man. Modern man does not. Or if he does, he should have his head examined. Modern libertarians will also balk at his puritanical rigidity on morals and stimulants, all the way down to and including a cup of tea. There are numerous other difficulties, including those "personal idiosyncracies." William Lloyd Garrison was a strange and difficult man.

Miss Kraditor appears to believe, however, that history has in a measure vindicated Garrison. "The policy of the 'realistic' political abolitionists [his opponents] did not, after all, produce peaceful abolition, and the alternatives that Garrison presented . . . were presented by life itself a score of years later." It did, after all, as she points out, take a revolution, a civil war, and a repudiation of the old Constitution, as well as some drastic shaking up of other institutions to abolish slavery. She goes further in a concluding passage "to speculate what the result would have been if a large part of the abolitionist movement had not weakened the moral force of its propaganda and accepted the compromises dictated by political expediency." She can not help wondering "whether the abolitionist movement did not yield too much when the major part of it, during the 1840's, played down the purely agitational sort of tactics in favor of political action that gave increasing emphasis to pragmatic alliances with politicians who would not denounce slavery in the abstract." She might have extended these speculations to ask whether those alliances and compromises did not burden the cause of emancipation so heavily with racist ideology and allies as to vitiate the cause itself. These are all the more interesting in view of a rich mine of recent scholarly investigation of white American attitudes toward race, race policy, and slavery in mid-nineteenth century America that make such speculation more profitable and informed than it might otherwise have been. Several years ago Leon Litwack opened up this field with a stimulating survey.[5] Later studies get down to the particulars and the in-fighting. Only at this level can one assess the wisdom and the chances of success that the various alternative policies might have had and test the claim of historical vindication for Garrison.

In the first place, while conceding the sincerity of abolitionist hatred of slavery and concern for the welfare of the Negro, several historians have recently pointed out that race prejudice of various kinds was endemic among the white abolitionists themselves. This tendency has been spelled out most explicitly by William H. Pease and Jane H. Pease,[6] who find that "antislavery crusaders were beset by a fundamental ambivalence in their attitude toward the Negro himself." Abolitionist pronouncements cited to illustrate this attitude range from the crudely explicit to subtly implicit and often unconscious stereotypes. Of the more explicit examples, Theodore Parker, a supporter of John Brown, could write in 1860 that "the Anglo-Saxon with common sense does not like this Africanization of America; he wishes the superior race to multiply rather than the inferior." More often the attitude betrayed a romanticized racial "stereotype of the malleable, willing and docile colored man." Thus, the antislavery rebels of Lane Seminary concluded in their de-

[5] Leon F. Litwack, *North of Slavery: The Negro in the Free States, 1790–1860* (Chicago, 1961).

[6] William H. Pease and Jane H. Pease, "Antislavery Ambivalence: Immediatism, Expediency, Race," *American Quarterly*, XVII (1965) 682–95. See also Litwack, *North of Slavery*, 216–30. Other historians such as Stanley Elkins, Larry Gara, and Louis Filler have referred to these attitudes among abolitionists.

bates that Negroes "would be kind and docile if immediately emancipated"; J. Miller McKim praised "their susceptibility to control"; and Angelina Grimké wrote a Negro friend with staggering tactlessness that "your long-continued afflictions and humiliations was the furnace in which He was purifying you from the dross, the tin, and the reprobate silver. . . ." Abolitionists measured the Negro by white, middle-class standards and expected him to live up to those standards. They were careful to disclaim approval of adopting colored children, encouraging interracial marriage, or "exciting the people of color to assume airs." They debated endlessly whether and to what degree to admit Negroes to their antislavery societies without ever wholly resolving their doubts. As Mr. and Mrs. Pease write:

> Never could the abolitionists decide collectively, and infrequently individually, whether the Negro was equal or inferior to the white; whether social equality for the Negro should be stressed or whether it should be damped; whether civil and social rights should be granted him at once or only in the indefinite and provisional future. . . . The abolitionists, furthermore, were torn between a genuine concern for the welfare and uplift of the Negro and a paternalism which was too often merely the patronizing of a superior class.

The ambivalence and temporizing of abolitionists is better understood in the light of an excellent study of antiabolition mobs by Leonard L. Richards.[7] With its help, we can begin to understand what abolitionists were up against and the willingness of all but a few to temporize and compromise. Employing ingenious techniques and imaginative scholarship, Mr. Richards has studied "all the major and minor mobs, riots, disturbances, civil disorders, and the like" reported between 1812 and 1849. More than half of them, 115 out of 207, occurred in the 1830's and were concentrated in the middle years of that decade, and 48 of the incidents in the 1830's and 1840's were antiabolitionist and racial in character. As Lincoln pointed out, mob violence

[7] Leonard L. Richards, *"Gentlemen of Property and Standing": Anti-Abolition Mobs in Jacksonian America* (New York, 1970).

had become a feature of American life. Great and small cities, small towns and rural communities in all parts of the country fell under the mercy of the mob, some of them for days at a time.

The conventional picture of mobs, antiabolition and anti-Negro mobs included, is that they were lower class, spontaneous, unorganized, and indiscriminate in their violence. This picture bears no resemblance to the typical antiabolition mobs Richards has carefully analyzed. "They were neither revolutionary nor lower-class. They involved a well-organized nucleus of respectable, middle-class citizens who wished to preserve the status quo rather than to change it. They met purposefully and often formally, and they coordinated their actions several days in advance. Frequently, they had either the support or the acquiescence of the dominant forces in the community. Sometimes they represented the Establishment. More frequently they *were* the Establishment." They were "usually led or engineered by the scions of old and socially dominant Northeastern families," by doctors and lawyers, merchants and bankers, judges and congressmen — "gentlemen of property and standing," in Garrison's own phrase. The collective wealth of mob members was appreciably greater than that of abolitionists of a given community. Yet subservience to Southern economic pressure was not a significant motive. "There were too many Northern mobs in cities and towns such as Montpelier, Utica, Lockport, Troy, and Granville — in places that had little or no dependence on Southern patronage." Fear and hatred of the black man played a part, to be sure, but "Negrophobia had been common for as long as anyone cared to remember," and blacks were no real threat to these disturbed gentlemen. Rather they saw the abolitionists as a symbol of all the dread forces of a new America that were challenging their elite status, the moral legitimacy of the established order, and threatening traditional means of social control. With their pressure groups, mass petitions, mass press, and free newspapers, the abolitionists, like so many other and even more powerful (but less vulnerable) subversive forces of centralization, were undermining the authority of local men of "character" and "respectability" who had dominated the old Northern Jacksonian society of gentlemen merchants. The abolitionists short-circuited these local elites, bypassed the city fathers, appealed over their heads directly to young and old, women as well as men, blacks as well as whites, to preach "amalgamation" and "leveling." The antiabolition mobs were in part a reaction against this challenge to traditional prerogatives of community leaders.

Abolitionist organizations and activities declined sharply after the Panic of 1837 cut back their funds, reduced their agents and publications, and broke up their unity. Simultaneously the number of antiabolitionist mobs declined sharply — the response with the challenge. Yet, paradoxically, antislavery converts then became more numerous and widespread. An important reason for this suggested by Mr. Richards was "that the ground upon which one might hold antislavery views was tremendously broadened in the late 1830's and 1840's." Abolitionism had become linked with other issues having little to do with its own validity — not only freedom of speech, press, and petition, but broadest of all with such issues as free soil. One could join the free soilers not only because he opposed slavery or its expansion, but also because he feared and hated Negroes. The antislavery leaders of the Free Soil Party in 1852 dropped the demand for equal rights for free Negroes of the North because, as Eric Foner has pointed out, they "realized that in a society characterized by all but universal belief in white supremacy, no political party could function effectively which included a call for equal rights [for blacks] in its national platform." [8] Under these conditions some of the bitterest antiabolitionists and even some of the mob leaders of the 1830's became prominent anti-

[8] Eric Foner, "Politics and Prejudice: The Free Soil Party and the Negro, 1849–1852," *Journal of Negro History,* L (October 1965), 239–40.

slavery politicians in the 1850's and Radical Republicans later on.[9] The movement that eventually democratized antislavery sentiment contrived to make it respectable at the Negro's expense. This was not Garrison's movement, for he abhorred racism as much as he did slavery. The successful antislavery movement embraced anti-Negro recruits and their prejudices and triumphed over slavery in the name of white supremacy.

In no part of the country was the combination of these sentiments and the alliance of those holding them more evident than in the West. It is the thesis of Eugene H. Berwanger's *The Frontier Against Slavery* that "prejudice against Negroes was a factor in the development of antislavery feeling in the antebellum United States."[10] His study "concentrates on the ever-shifting frontier regions which became free states or territories by 1860" but which were threatened at some time with the legalization of slavery. This latest investigation of "frontier influences" offers cold comfort to Frederick Jackson Turner's hypothesis that the frontier experience was a major source and an ever-rejuvenating stimulus of the democratic impulse in American life — if democracy and American life included Negro people. Without explicitly making the comparisons suggested, Berwanger lends less support to Turner's thesis than to Tocqueville's observation that "the prejudice of race appears to be stronger in the states which have abolished slavery than in those where it still exists; and nowhere is it so important as in those states where servitude never has been known."[11]

Proslavery men kept up their fight for legalizing bondage in

[9] For example, John Parker Hale of Dover, New Hampshire, disrupted and broke up a series of antislavery lectures in 1835 by a speech in which he declared that Negro slaves *"are beasts in human shape, and not fit to live,"* adding in a barely audible voice, "free." Hale later became the standard bearer of the Liberty party in 1847 and the presidential candidate of the Free Soil party in 1852. Richards, *"Gentlemen of Property and Standing,"* 163–64.

[10] Eugene H. Berwanger, *The Frontier Against Slavery: Western Anti-Negro Prejudice and the Slavery Extension Controversy* (Urbana, 1967).

[11] Alexis de Tocqueville, *Democracy in America* (New York, 1904), I, 383.

the Old Northwest until the 1830's, but by 1811 antislavery and anti-Negro forces of the Indiana territorial legislature had passed laws preventing Negroes from testifying in court against whites, excluding them from militia duty, and barring them from voting. Ohio in 1807 excluded Negroes from residence in the state unless they posted a $500 bond for good behavior. In 1813, Illinois ordered every incoming free Negro to leave the territory under a penalty of 39 lashes repeated every fifteen days until he left. The chief argument against slavery was that it would eventually produce a free Negro population. By the early 1830's all three states had adopted "almost identical statute restrictions against free Negroes." Michigan and Iowa followed suit in their turn, and Wisconsin joined them in voting down Negro suffrage by large majorities. Anti-Negro activity intensified in the 1840's and 1850's, when Illinois and Indiana wrote harsh free-Negro-exclusion provisions into their constitutions. John A. Logan of Illinois, future radical Republican leader, was the author of "the most severe anti-Negro measure passed by a free state."

The peak of the anti-Negro movement in the Middle West coincided with the height of agitation over the slavery-expansion issue, but Mr. Berwanger's evidence indicates that often the main concern was not so much over the expansion of slavery as the migration of Negroes. David Wilmot, author of the famous "Proviso" excluding slavery from new territories, avowed his object was to "preserve to free labor . . . of my own race and own color" those new lands. Horace Greeley declared that the unoccupied West "shall be reserved for the benefit of the white Caucasian race." Owen Lovejoy, congressman from the "most abolitionized" district of Illinois and brother of the martyred abolitionist Elijah, denounced the idea of Negro equality. Joshua Giddings, zealous antislavery congressman from Ohio, pronounced Negroes "not the equal of white men," and Benjamin F. Wade, another congressman of the same state and same persuasion called Washington "a Nigger-ridden place." Lyman Trumbull, Illinois Republican leader and a close friend of Lincoln, declared that "We, the Republican Party, are a white man's party."

The region that inspired these obsessions and phobias against Negro people never had a black population in this era exceeding one percent of the total. Yet the attitudes and laws developed in the Middle West became the models for new territories farther westward as Midwesterners followed the frontier to the Pacific. As Mr. Berwanger says, "These pioneers pushed westward with an increased determination to keep the Negro, free or slave, out of the new lands."

Before 1848, Californians under Mexican rule are said to have "accepted Negroes as equal individuals" and intermarried with them, but this situation quickly changed with the arrival of great numbers of Americans the following year. Delegates to the Constitutional Convention of California in 1849 voted without opposition and without debate to adopt the same constitutional restrictions on free Negroes in their fundamental laws that were found in the Middle West. A measure excluding free Negroes from the territory entirely was defeated only because of fear that this would delay congressional action on statehood. Californians voted down slavery, but an informed citizen said that "not one in ten cares a button for its abolition . . . all they look at is their own position; they must themselves swing the pick, and they won't swing it by the side of Negro slaves." The state legislature swiftly debarred blacks from testifying against whites in court and from intermarriage with whites, while local authorities segregated them. In 1857, the state prison director shipped black inmates of the prisons to New Orleans, where they were sold into slavery.

The Pacific Northwest in its turn duplicated the laws and racial customs of the Old Northwest and was settled in considerable measure from that region. The territorial legislature of Oregon subjected Negroes who refused to leave to periodic floggings and later substituted apprenticeship to white men. The constitutional convention for statehood in 1857 rapidly approved laws excluding Negroes from the militia and the polls. A popular referendum rejected slavery by a majority of 5,000, but approved exclusion of free Negroes from the state by a majority of 7,500 — 8,640 to 1,081. "Slavery and the Negro," it seems, "to the average farmer, were one and the same." Congress admitted Oregon in 1857, the only free state with a Negro exclusion clause in its original constitution ever admitted. "Oregon is a land for the white man," said the *Oregon Weekly Times*. "Refusing the toleration of Negroes in our midst as slaves, we rightly and for yet stronger reasons, prohibit them from coming among us as free Negro vagabonds."

"Bleeding Kansas" was the *cause célèbre* among Eastern antislavery radicals in the mid-1850's, but a census of 1856 showed that 83 percent of Kansas settlers had lived in the Old Northwest, or in Iowa, Kentucky, or Missouri. Mr. Berwanger proves that "a large group of settlers was more anti-Negro than antislavery." The population was, of course, divided politically and sometimes militarily between the proslavery and antislavery factions. But even among the latter, except for a minority in the area of Lawrence, reliable evidence shows that "anti-Negro sentiment was overwhelming among the antislavery settlers." A referendum which the proslavery people boycotted and which was confined to the antislavery population polled 1,287 for and 453 against excluding free Negroes from the territory. Thus three out of every four antislavery Kansans approved Negro exclusion. The antislavery heroes of the Western plains to whom Boston abolitionists were shipping "Beecher's Bibles" were at that time engaged in limiting suffrage, office holding, and militia service to white men. They were evidently more anti-Negro than they were antislavery. The same was true of Utah, Colorado, New Mexico, and Nebraska, which repeated the old story of discrimination, but with the added absurdity of having so few Negroes in their borders about whom to quarrel. With only 59 live Negroes in their territory to prove it, Mormons resolved that the race was inferior to whites even "in the next world." Nebraska had only 67 Negroes in 1860, yet Negro exclusion had been a hot issue in 1859.

The modesty and thoroughness of Mr. Berwanger's scholarship is indicated by his admission that "the exact extent of racial prejudice as a factor encouraging the limitation of slavery is indeterminable." But he adds cogently that "if 79.5 percent of the people of Illinois, Indiana, Oregon and Kansas voted to exclude the free Negro simply because of their prejudice, surely this antipathy influenced their decisions to support the nonextension of slavery." And surely any fair-minded reader will concede that he has a point there.

All writers on Midwestern attitudes toward slavery and the Negro point out that these attitudes may be partially accounted for by the considerable migration of nonslaveholding Southern yeomen into some parts of that region. What then of the attitudes in the Middle Atlantic and New England states, where the number of Southern migrants was negligible? In a less ambitious study than the previous one, Lorman Ratner treats the Northern states omitted by Berwanger — those of the Northeast.[12]

It is not difficult to establish that, in the period concerned, the great majority of people in the Northeast despised or opposed abolitionists and were hostile or indifferent to the antislavery cause. Mr. Ratner undertakes to go further, however, and explain why they had those attitudes. What he actually does is to tell us why they *said* they opposed the movement. What they said is of interest even if it does not explain. For example, two great theologians, Charles Grandison Finney and William Ellery Channing, and Noah Porter, later President of Yale, agreed that abolishing slavery would aggravate the race problem. Two powerful clergymen, Lyman Beecher and Horace Bushnell, were afraid that it would place Negroes on an equal basis with whites. James Fenimore Cooper, the novelist, and Francis Wayland, clergyman and authority on ethics, feared that free Negroes would be the prey of demagogues and subvert the state. Seth Luther and George Henry Evans, labor spokesmen, feared that abolition would result in unfair competition for white labor.

[12] Lorman Ratner, *Powder Keg: Northern Opposition to the Antislavery Movement, 1831–1840* (New York, 1968).

Everybody had reasons, all kinds of reasons, logical and frivolous reasons, informed and superficial reasons. Among churchmen, the Presbyterians, Congregationalists, Baptists, Methodists, Episcopalians, Lutherans, Unitarians, Catholics all took positions at one time "that placed the majority of their clerical leaders in opposition to the antislavery movement." This does not mean that these churches were necessarily dominated by proslavery men. Indeed, opponents of abolition were often opponents of slavery as well. They might deplore slavery but deplore abolition more and abolitionists even more than that. The commonest reason given by churchmen when speaking for their churches was that abolition would disrupt their institutions by dividing the members — which it undoubtedly would in most instances.

All the Northeastern states had abolished or taken steps for the gradual abolition of slavery within their own borders by 1804, but racial discrimination, segregation, and injustice still flourished and often proliferated in those regions. Only the five states with the most insignificant numbers of blacks, all in New England, permitted them to vote. New York had property qualifications that withheld the ballot from nearly all of them. Massachusetts outlawed intermarriage and for a while tightened segregation, and the other states followed much the same policy. The Pennsylvania legislature formally refused the Negro the vote in 1837 and at the same time seriously debated, though did not pass, an exclusion bill of the Western type to prevent free Negroes from entering the state. Northeasterners of high and low status flaunted race prejudice openly. Negroes lived a thoroughly segregated existence, set apart in church, school, public services, and society, excluded from politics and handicapped in the courts of justice. They constituted, with few exceptions, a despised and ostracised caste.

Mr. Ratner is probably right that the "romantic image of the South" created by Northern novelists[13] fostered "northern re-

[13] Analyzed by William R. Taylor, *Yankee and Cavalier: The Old South and American National Character* (New York, 1961).

spect and admiration, even pride in that region." This attitude toward the South doubtless affected Northern attitudes toward slavery and the Negro. As the sectional crisis intensified in the 1850's, however, the romantic image faded and the South, not the Negro, came to be regarded as the real menace to the North. Yet the North entered the Civil War as part of a slave republic to defend a Constitution that guaranteed slave property, led by a party with a platform pledged to protect slavery where it existed, and headed by a President who declared in his inaugural address that he had "no lawful right" and "no inclination" to interfere with slavery in the South. Congress followed this up three months after the war started with virtually unanimous resolutions viewing with horror any suggestion of permitting the war to interfere with "established institutions of those States."

The party that led the Northern crusade against slavery has not come off well at the hands of the new revisionists who have been emphasizing the pervasiveness of anti-Negro sentiment in Northern society. Not only the conservative and the moderate, the ex-Whig and ex-Democratic Republicans, but the radical, antislavery, and abolitionist Republicans as well have been found wanting in their racial attitudes and policies. They have been caught mouthing the slogans of white privilege and white supremacy, appealing to race prejudices of the electorate, appeasing mass phobias, and condoning compromises with discriminatory and outrageously unjust anti-Negro policies and laws. This derogatory emphasis on the seamy side and the hypocritical aspects of the Republican party, especially the Radicals, is all the more striking because it follows hard upon a decade of historiography devoted to the rehabilitation and vindication of the Radical Republicans. In the hands of these historians, what had probably been until about 1960 "the best-hated men in American history" were on the way to becoming the most elaborately vindicated and widely venerated politicians in our history.[14] It was to be expected, therefore, that the new counterin-

[14] Hans L. Trefousse, *The Radical Republicans: Lincoln's Vanguard for Racial Justice* (New York, 1969), attempts "to draw together the

terpretation emphasizing the retrograde and racist side of Radical Republicans would not long go unchallenged.

The first challenge of consequence comes from the pen of Eric Foner in a work on Republican ideology before the Civil War.[15] Conciliatory and low-keyed, his response is free of the sweeping tone of vindication and unqualified defense. Admitting that "racism and colonization were important elements of the Republican attitude toward the Negro," he contends that "they were by no means the entire story," and that those who take Republicans to task for racial prejudice "have carried a good point too far." He points out, for example, that even in the Middle West, enclaves of radical Republicanism, such as fourteen counties of northern Illinois and the Western Reserve of Ohio, consistently opposed black laws and exclusion laws. He believes that "a majority of Republicans [in the West] were ready to give some recognition to Negro rights" and that in Ohio "some Republican legislators favored Negro suffrage." Mr. Foner is able to list Republicans of high rank who advocated political rights for free Negroes, opposed exclusion laws, and defended and assisted fugitive slaves, and a few who fought for Negro suffrage. This alone, he thinks, "should be proof that there was more to Republican attitudes than mere racism." Out of experience and debate emerged "what may be called the mainstream of Republican opinion" on Negro rights. "Fundamentally it asserted that free Negroes were human beings and citizens" and deserved protection of life, liberty, and property. Although these

findings" of this school. While he admits that there was a "diversity of approaches to the race question" among radicals and cites evidence of their racial prejudices, his conclusions about the radicals are better reflected in the subtitle of his book. "The villains of yesteryear," he observes, "are the heroes of today." A briefer synthesis is Larry Kinkaid, "Victims of Circumstance: An Interpretation of Changing Attitudes toward Republican Policy Makers and Reconstruction," *Journal of American History,* LVII (June 1970), 48–66. The assessment covers the historiography of vindication, but not the later literature that runs counter to it.

[15] Eric Foner, *Free Soil, Free Labor, Free Men: The Ideology of the Republican Party Before the Civil War* (New York, 1970).

rights were not extended to slaves, did not include equality or the right to vote and hold office, and were "not accepted by everyone in the party," they constituted "a distinctive Republican position." And in view of pervasive Northern racism, "the Republicans' insistence on the humanity of the Negro was more of a step forward than might appear."

This is not a very strong case for the defense, and was not intended to be. It relies heavily on the point that the Democrats were even worse, which is true. "To a large extent, these [Republican] expressions of racism were political replies to Democratic accusations rather than gratuitous insults to the black race." The case is further weakened by the admission that "inherent in the antislavery outlook of many Republicans was a strong overtone of racism" and that the party borrowed much from the "free labor" position that abhorred the black man about as much as the slave. It is further conceded that "no portion of the Republican party could claim total freedom" from racism. "Nevertheless," contends Mr. Foner, "by the eve of the Civil War there had emerged a distinctive attitude toward the Negro." This was best represented by Abraham Lincoln, who acknowledged the "universal feeling" against the Negro in the North and shared the opposition to Negro suffrage and Negro political and social equality, but who insisted on "the basic humanity of the Negro," his right to earn a livelihood and to have protection of his basic rights.

In spite of the Republican party's formal disavowal of any intention to interfere with slavery, within two years after the opening of the Civil War the Union was formally committed to freedom as a war aim and later in some degree (so the present historian once maintained) to the far more revolutionary war aim of racial equality. By these commitments, a power struggle with the negative aim of preventing secession was metamorphosed into a war of ideologies, a moral crusade with divinely sanctioned ends. Exalted by the crusading spirit, many Union participants, especially in retrospect, hallowed their cause with a "common glory" and sanctified it as a holy war. Thus the American Civil War entered into legend and history with an aura of sanctity that few wars in history have enjoyed. The extent to which Union policies and attitudes toward the Negro justified this view of the war is subject to searching investigation by V. Jacque Voegeli.[16] He focuses his study on the Middle West, but his findings have wider import and deserve respectful attention.

Convincing evidence supports Voegeli's conclusion that the outbreak of the Civil War actually "increased the virulence of midwestern racism." For the war opened the prospect of an inundation of the North by fugitive or liberated Negroes. Thus the fears that motivated so much opposition to the *expansion* of slavery were now revived and intensified by the threatened *abolition* of slavery. Congressmen and editors gave full voice to these fears of "free Negroism," or "Africanization," fears of an invasion of blacks who would "fill the jails and poorhouses, compete with white labor, and degrade society." Ohioans were not prepared to "mix up four millions of blacks with their sons and daughters." The *Chicago Times* predicted that emancipation would flood the North with "two or three million semi-savages." Lincoln's own state, where the constitution and statutes already barred further Negro settlement, endorsed new Negro exclusion measures in a referendum of June 1862, by a vote of two to one.

The urgent question for the politicians was framed by a correspondent of Senator Ben Wade: "If we are to have no more slave states what the devil are we to do with the surplus niggers?" Lyman Trumbull, speaking for the West as a whole and for Illinois in particular, told the Senate, "Our people want nothing to do with the Negro." Salmon P. Chase urged General Ben Butler "to see that the blacks of the North will slide Southward, and leave no question to quarrel about." Others held out the hope that after slavery was abolished freedmen would want to stay in the South and Northern Negroes would actually prefer to live there. Hoping to placate Northern fears, Republicans made

[16] V. Jacque Voegeli, *Free but Not Equal: The Midwest and the Negro During the Civil War* (Chicago, 1967).

Negro deportation and colonization abroad the official policy of the party, and "openly avowed in Congress that deportation was designed partly to keep the freedmen out of the North." President Lincoln was a strong advocate of the policy. In view of the keen instinct for the practical and possible that Lincoln had, it is hard to believe his advocacy was other than political in motivation.

The backlash to emancipation was frightening. Lincoln thought his Proclamation had "done about as much harm as good." The Illinois legislature adopted a resolution denouncing it and debated enforcement of the state exclusion law with "thirty-nine lashes on the bare back." Anti-Negro violence erupted in Toledo, Chicago, Peoria and Cincinnati. At New Albany, Indiana, guards were stationed at ferries to prevent Negroes landing and a regiment of Indiana troops threatened to fire on Negroes and keep them from crossing the Ohio. Republican party leaders repeatedly disavowed intentions of permitting or encouraging Negro migration or tolerating Negro equality. The charge that "our volunteers are periling their lives to make the niggers equal" was denounced by the Indianapolis *Journal* as "a monstrous and villainous lie."

Deeds rather than words were required of the Republicans. The old clichés and promises were not enough. Everyone knew that hundreds of thousands of idle and helpless refugee freedmen were piling up behind the Union lines in the South. Secretary of War Edwin M. Stanton, seeing the need for labor in the North, blundered into shipping hundreds of blacks into Illinois, and then beat a hasty retreat when his action touched off a violent explosion of resentment. Lincoln's experiment with deportation and colonization of blacks in the Caribbean proved a failure. He began arming Negroes in spite of his "fear that in a few weeks the arms would be in the hands of the rebels," but the army could absorb only a fraction of the refugees. In March 1863, Lincoln placed the problem in the hands of General Lorenzo Thomas, a rather hapless old functionary. Speaking to Union troops in Louisiana the following month, "with full authority from the President,"

Thomas said of the refugees, "They are coming in upon us in such numbers that some provision must be made for them. You cannot send them North. You all know the prejudices of the Northern people. . . ." The policy decided upon, said the General, was deliberately to contain the Negroes in the South, put them to work on "the multitude of deserted plantations" he pointed to "along the river." The hope was to make them self-supporting, but at any rate to keep them down South. Mr. Voegeli concludes that the policy "effectually sealed the vast majority of them in the region," and that "the political motive was crucial" in the determination of the containment policy.

Northern racial phobias and "apprehensions of a black invasion" cooled somewhat after the policy of containment was publicized and its effectiveness became apparent. War service of the Negroes, their demands for justice, and their horrible persecution in the New York City riots gained some sympathy for them. The war then became "a limited crusade" in the eyes of many, "a struggle for God, mankind, liberty and Union." Combining "a mixture of idealism and vengefulness, innocence and arrogance" the limited crusaders saw the war as "a battle for cultural supremacy . . . the final confrontation of Puritan and Cavalier," in which "Southern vices . . . were to be supplanted by Northern virtues — industrialism, democracy, equality, prosperity, ingenuity, intelligence, and unselfish nationalism."

Several years ago, the present writer advanced the thesis that in this crusade phase a "third war aim," the boldly revolutionary aim of racial equality, was almost surreptitiously added by the radicals to the primary war aims of Union and Freedom. It was a qualified suggestion, later retracted.[17] Mr. Voegeli effectively attacks the original suggestion. I think he is right and I was wrong. Of course, some influential Northerners did want to make equality a war aim, and the postwar civil rights acts and

[17] The thesis of "the third war aim" was suggested by the author in *The Burden of Southern History* (Baton Rouge, 1960), and retracted in a second edition of that work (p. 87) published in 1968.

constitutional amendments can be read as evidence of their success. But as I said at the time, "legal commitments overreached moral persuasion," and the Union "fought the war on borrowed moral capital" and then repudiated the debt. It was nevertheless misleading to equate Equality with the commitments to Union and Freedom, and Mr. Voegeli is right in saying "there is no reason to believe that full equality for Negroes ever became one of their war aims." He is also right that there was "some bold talk . . . but little action," that "only a few of the innumerable extralegal devices which drew the color line in the Midwest were discarded during the war," that "on the whole, these changes scarcely scratched the surface of white supremacy," and that "in most places Negroes remained fundamentally as before — victims of discrimination . . . of social ostracism, and of economic subordination."

Sincerity of Union purpose may also be tested by federal race policy in the South after the war, and on this subject there is a revisionary work of superior quality by William S. McFeely.[18] It soon became apparent that Radical Reconstruction did not mean abandonment of the containment policy. Voegeli pointed out that an amendment to the Freedmen's Bureau bill sensibly providing for the organized dispersion of freedmen to jobs in the North was defeated in the Senate, where Senator Charles Sumner of Massachusetts called it "entirely untenable," and Senator Henry Wilson of the same state feared it would have "a bad influence." General Howard's mission was to solve the problem of freedmen within the South. All agreed then and still agree now that the one-armed "Christian General" was by all odds the best man for the job. He was given wide powers and a bill authorizing him to distribute "abandoned and confiscated land" to freedmen.

The freedmen never got the promised land to have and hold. It was taken away from them and returned to the planters, and

[18] William S. McFeely, *Yankee Stepfather: A Study of General O. O. Howard and the Freedmen's Bureau* (New Haven, 1968).

General Howard painfully and personally presided over the restitution. It is one of the stranger ironies of historiography that for a hundred years the Freedmen's Bureau has been pictured by friend and foe alike as an instrument of radical policy. Actually it was skillfully employed by President Andrew Johnson with the compliance of Howard to subvert radical purposes and advance conservative ends. On the complaint by influential whites, the General eventually removed virtually every subordinate who sought to fulfill the original mission of the Bureau and help the freedmen. "The Freedmen's Bureau had not stopped the delivery of the Negro labor force into the hands of Johnson's planters and businessmen allies," writes Mr. McFeely. "On the contrary, it was used by the President to accomplish this purpose." In less skillful hands, this important revision of history might have taken the tone of strident cynicism and exposure. Fortunately Mr. McFeely has the subtlety and grace to make his story "not an exposé of a knave, but rather a record of naïveté and misunderstanding." Thus he is able not only to grasp the tragedy of the freedmen, but the tragedy of General Howard as well, and to see both as the failure of America.

The findings of the revisionists might help with Miss Kraditor's speculations about history vindicating Garrison and what might have happened had abolitionists remained faithful to his ideals. History supports Garrisonian dogma that "reform" was not enough, that to be effective the eradication of slavery had to be root-and-branch, that the racist ideology supporting it permeated the country, and that abolishing slavery in alliance with racists and without eradicating their ideology would be largely an empty victory. To grant that agitation is "the art of the desirable" is to concede Garrison much. But when Garrison put aside his pacifism and many of his principles to support the Union cause and the Constitution on the ground that "death and hell had seceded," he shifted from agitation to politics, "the art of the possible." Of that art he was wholly innocent. The greatest artist of the possible in his time was now in charge, and Lincoln knew in his bones that unless he fought that war with the support of racists and in the name of white supremacy it would be *his* lost cause and not Jefferson Davis's.

To return finally to the questions raised at the beginning of this chapter, some of the answers still seem rather elusive and ambiguous. After an excursion through revisionary historical literature, however, it seems harder than ever to locate precisely that legendary "interlude of virtue" when Americans renounced their racial prejudices and rededicated themselves to their ideals of equality. The present seems depressingly continuous with the past. And now that the old policy of "containment" no longer holds, now that the fears of an inundation by Negro migrants from the South that obsessed the North of the 1860's have materialized a century later, the continuity of plantation and ghetto is borne in upon us. It also seems rather more difficult than it was before to be confident in justifying the sacrifice of those 600,000 lives. It was a crusade to be sure, but a "limited" one, and it does appear to have been corrupted and frustrated all along by an old sickness endemic among the crusaders.

6

Seeds of Failure in Radical Race Policy

THE Republican leaders were quite aware in 1865 that the issue of Negro status and rights was closely connected with the two other great issues of Reconstruction — who should reconstruct the South and who should govern the country. But while they were agreed on the two latter issues, they were not agreed on the third. They were increasingly conscious that in order to reconstruct the South along the lines they planned they would require the support and the votes of the freedmen. And it was apparent to some that once the reconstructed states were restored to the Union, the Republicans would need the votes of the freedmen to retain control over the national government. While they could agree on this much, they were far from agreeing on the status, the rights, the equality, or the future of the Negro.

The fact was that the constituency on which the Republican congressmen relied in the North lived in a race-conscious, segregated society devoted to the doctrine of white supremacy and Negro inferiority. "In virtually every phase of existence," writes Leon Litwack with regard to the North in 1860, "Negroes found themselves systematically separated from whites. They were either excluded from railway cars, omnibuses, stage coaches, and steamboats or assigned to special 'Jim Crow' sections; they sat, when permitted, in secluded and remote corners of theatres and lecture halls; they could not enter most hotels, restaurants, and resorts, except as servants; they prayed in 'Negro pews' in the white churches. . . . Moreover, they were often educated in segregated schools, punished in segregated prisons, nursed in segregated hospitals, and buried in segregated cemeteries." Ninety-four percent of the Northern Negroes in 1860 lived in states that denied them the ballot, and the six percent who lived in the five states that permitted them to vote were often disfranchised by ruse. In many Northern states, discriminatory laws excluded Negroes from interracial marriage, from militia service, from the jury box, and from the witness stand when whites were involved. Ohio denied them poor relief, and Indiana, Illinois, and Iowa had laws carrying severe penalties against Negroes settling in those states. Everywhere in the free states, the Negro met with barriers to job opportunities and in most places he encountered severe limitations to the protection of his life, liberty, and property.[1]

One political consequence of these racial attitudes was that the major parties vied with each other in their professions of devotion to the dogma of white supremacy. Republicans were especially sensitive on the point because of their antislavery associations. Many of them, like Senator Lyman Trumbull of Illinois, found no difficulty in reconciling antislavery with anti-Negro views. "We are for free white men," said Senator Trumbull in 1858, "and for making white labor respectable and honorable, which it can never be when negro slave labor is brought into competition with it." Horace Greeley the following year regretted that it was "the controlling idea" of some of his fellow Republicans "to prove themselves 'the white man's party,' or else all the mean, low, ignorant, drunken, brutish whites will go against them from horror of 'negro equality.' " Greeley called such people "the one-horse politicians," but he could hardly apply that name to Lyman Trumbull, nor for that matter to William H. Seward, who in 1860 described the American Negro as

[1] Leon Litwack, *North of Slavery: The Negro in the Free States, 1790–1860* (Chicago, 1961), 91–97.

"a foreign and feeble element like the Indians, incapable of assimilation," nor to Senator Henry Wilson of Massachusetts, who firmly disavowed any belief "in the mental or the intellectual equality of the African race with this proud and domineering white race of ours." [2] Trumbull, Seward, and Wilson were the front rank of Republican leadership and they spoke the mind of the Middle West, the Middle Atlantic states, and New England. There is much evidence to sustain the estimate of W. E. B. Du Bois that "At the beginning of the Civil War probably not one white American in a hundred believed that Negroes could become an integral part of American democracy." [3]

When the war for Union began to take on the character of a war for Freedom, Northern attitudes toward the Negro, as demonstrated in the previous chapter, paradoxically began to harden rather than soften. This hardening process was especially prominent in the Middle Western states where the old fear of Negro invasion was intensified by apprehensions that once the millions of slaves below the Ohio River were freed they would push northward — this time by the thousands and tens of thousands, perhaps in mass exodus, instead of in driblets of one or two who came furtively as fugitive slaves. The prospect filled the whites with alarm and their spokesmen voiced these fears with great candor. "There is," Lyman Trumbull told the Senate, in April 1862, "a very great aversion in the West — I know it to be so in my state — against having free negroes come among us." [4] And about the same time, John Sherman, who was to give his name to the Radical Reconstruction acts five years later, told Congress that in Ohio "we do not like negroes. We do not disguise our dislike. As my friend from Indiana [Congressman Joseph A. Wright] said yesterday, the whole people of the northwestern

[2] Quoted in *ibid.*, 92, 269–272.

[3] W. E. B. Du Bois, *Black Reconstruction in America, 1860–1880* (New York, 1935), 191.

[4] Quoted in Jacque Voegeli, "The Northwest and the Race Issue, 1861–1862," *Mississippi Valley Historical Review*, L (1963), 240.

States are, for reasons whether correct or not, opposed to having many negroes among them and the principle or prejudice has been engrafted in the legislation of nearly all the northwestern States." [5]

So powerful was this anti-Negro feeling that it almost overwhelmed antislavery feeling and seriously imperiled the passage of various confiscation and emancipation laws designed to free the slave. To combat the opposition Republican leaders such as George W. Julian of Indiana, Albert G. Riddle of Ohio, and Treasury Secretary Salmon P. Chase advanced the theory that emancipation would actually solve Northern race problems. Instead of starting a mass migration of freedmen northward, they argued, the abolition of slavery would not only put a stop to the entry of fugitive slaves but would drain the Northern Negroes back to the South. Once slavery were ended, the Negro would flee Northern race prejudice and return to his natural environment and the congenial climate of the South. [6]

The official answer of the Republican party to the Northern fear of Negro invasion, however, was deportation of the freedmen and colonization abroad. The scheme ran into opposition from some Republicans, especially in New England, on the ground that it was inhumane as well as impractical. But with the powerful backing of President Lincoln and the support of Western Republicans, Congress overcame the opposition. Lincoln was committed to colonization not only as a solution to the race problem but as a means of allaying Northern opposition to emancipation and fears of Negro exodus. To dramatize his solution, the President took the unprecedented step of calling Negro leaders to the White House and addressing them on the subject. "There is an unwillingness on the part of our people," he told them on August 14, 1862, "harsh as it may be, for you free colored people to remain with us." He told them that "your race suffer very greatly, many of them by living among us, while ours

[5] *Congressional Globe*, 37 Cong., 2 Sess. (April 2, 1862), 1495.
[6] Voegeli, 240–41.

suffer from your presence. . . . If this be admitted, it affords a reason at least why we should be separated." [7]

The fall elections following the announcement of the Emancipation Proclamation were disastrous for the Republican party. And in his annual message in December the President returned to the theme of Northern fears and deportation. "But it is dreaded that the freed people will swarm forth and cover the whole land?" he asked. They would flee the South, he suggested, only if they had something to flee from. *"Heretofore,"* he pointed out, "colored people to some extent have fled North from bondage; and *now,* perhaps, from both bondage and destitution. But if gradual emancipation and deportation be adopted, they will have neither to flee from." They would cheerfully work for wages under their old masters "till new homes can be found for them in congenial climes and with people of their own blood and race." But even if this did not keep the Negroes out of the North, Lincoln asked, "in any event, can not the north decide for itself, whether to receive them?" [8] Here the President was suggesting that the Northern states might resort to laws such as several of them used before the war to keep Negroes out.

During the last two years of the war Northern states began to modify or repeal some of their anti-Negro and discriminatory laws. But the party that emerged triumphant from the crusade to save the Union and free the slave was not in the best political and moral position to expand the rights and assure the equality of the freedman. It is difficult to identify any dominant organization of so-called "Radical Republicans" who were dedicated to the establishment of Negro equality and agreed on a program to accomplish their end. Both Southern conservatives and Northern liberals have long insisted or assumed that such an organization of radicals existed and determinedly pursued their purpose. But the evidence does not seem to support this assumption. There undoubtedly *did* emerge eventually an organization deter-

[7] Roy P. Bassler (ed.), *The Collected Works of Abraham Lincoln* (9 vols., New Brunswick, 1953), V, 371–72. [8] *Ibid.*, 535–36.

mined to overthrow Johnson's policies and take over the control of the South. But that was a different matter. On the issue of Negro equality the party remained divided, hesitant, and unsure of its purpose. The historic commitment to equality it eventually made was lacking in clarity, ambivalent in purpose, and capable of numerous interpretations. Needless to say, its meaning has been debated from that day to this.

The Northern electorate that the Republicans faced in seeking support for their program of reconstruction had undergone no fundamental conversion in its wartime racial prejudices and dogmas. As George W. Julian told his Indiana constituents in 1865, "the real trouble is that *we hate the negro*. It is not his ignorance that offends us, but his color." [9]

In the years immediately following the war every Northern state in which the electorate was given the opportunity to express its views on issues involving racial relations reaffirmed, usually with overwhelming majorities, its earlier and conservative stand. This included the states that reconsidered — and reaffirmed — their laws excluding Negroes from the polls, and others that voted on such questions as office holding, jury service, and school attendance. Throughout these years, the North remained fundamentally what it was before — a society organized upon assumptions of racial privilege and segregation. As Senator Henry Wilson of Massachusetts told his colleagues in 1867, "There is today not a square mile in the United States where the advocacy of the equal rights of those colored men has not been in the past and is not now unpopular." [10] Whether the Senator was entirely accurate in his estimate of white opinion or not, he faithfully reflects the political constraints and assumptions under which his party operated as they cautiously and hesitantly framed legislation for Negro civil and political rights — a program they knew had to be made acceptable to the electorate that Senator Wilson described.

[9] George W. Julian, *Speeches on Political Questions* (New York, 1872), 299.

[10] *Congressional Globe,* 40 Cong., 3 Sess. (Jan. 28, 1869), 672.

This is not to suggest that there was not widespread and sincere concern in the North for the terrible condition of the freedmen in the South. There can be no doubt that many Northern people were deeply moved by the reports of atrocities, peonage, brutality, lynchings, riots, and injustices that filled the press. Indignation was especially strong over the Black Codes adopted by some of the Johnsonian state legislatures, for they blatantly advertised the intention of some Southerners to substitute a degrading peonage for slavery and make a mockery of the moral fruits of Northern victory. What is sometimes overlooked in analyzing Northern response to the Negro's plight is the continued apprehension over the threat of a massive Negro invasion of the North. The panicky fear that this might be precipitated by emancipation had been allayed in 1862 by the promises of President Lincoln and other Republican spokesmen that once slavery was abolished, the freedmen would cheerfully settle down to remain in the South, that Northern Negroes would be drawn back to the South, and that deportation and colonization abroad would take care of any threat of Northern invasion that remained. But not only had experiments with deportation come to grief, but Southern white persecution and abuse combined with the ugly Black Codes had produced new and powerful incentives for a Negro exodus while removal of the shackles of slavery cleared the way for emigration.

The response of the Republican Congress to this situation was the Civil Rights Act of 1866, later incorporated into the Fourteenth Amendment. Undoubtedly part of the motivation for this legislation was a humanitarian concern for the protection of the Negro in the South, but another part of the motivation was less philanthropic and it was concerned not with the protection of the black man in the South but the white man in the North. Senator Roscoe Conkling of New York, a member of the Joint Committee of Fifteen who helped draft the Civil Rights provisions, was quite explicit on this point. "Four years ago," he said in the campaign of 1866, "mobs were raised, passions were roused, votes were given, upon the idea that emancipated negroes were to burst in hordes upon the North. We then said, give them liberty and rights at the South, and they will stay there and never come into a cold climate to die. We say so still, and we we want them let alone, and that is one thing that this part of the amendment is for." [11]

Another prominent member of the Joint Committee who had a right to speak authoritatively of the meaning of its racial policy was George Boutwell of Massachusetts. Addressing his colleagues in 1866, Boutwell said: "I bid the people, the working people of the North, the men who are struggling for subsistence, to beware of the day when the southern freedmen shall swarm over the borders in quest of those rights which should be secured to them in their native states. A just policy on our part leaves the black man in the South where he will soon become prosperous and happy. An unjust policy in the South forces him from home and into those states where his rights will be protected, to the injury of the black man and the white man both of the North and the South. Justice and expediency are united in indissoluble bonds, and the men of the North cannot be unjust to the former slaves without themselves suffering the bitter penalty of transgression." [12] The "bitter penalty" to which Boutwell referred was not the pangs of a Puritan conscience. It was an invasion of Southern Negroes. "Justice and expediency" were, in the words of a more famous statesman of Massachusetts, "one and inseparable."

The author and sponsor of the Civil Rights Act of 1866 was Senator Lyman Trumbull, the same man who had in 1858 described the Republicans as "the white man's party," and in 1862 had declared that "our people want nothing to do with the negro." Trumbull's bill was passed and, after Johnson's veto, was repassed by an overwhelming majority. Limited in application,

[11] Alfred R. Conkling, *The Life and Letters of Alfred R. Conkling, Orator, Statesman, Advocate* (New York, 1889), 277.

[12] Quoted in Benjamin B. Kendrick (ed.), *The Journal of the Joint Committee of Fifteen on Reconstruction* (New York, 1914), 341–42.

the Civil Rights Act did not confer political rights or the franchise on the freedmen.

The Fourteenth Amendment, which followed, was even more equivocal and less forthright on racial questions and freedmen's rights. Rejecting Senator Sumner's plea for a guarantee of Negro suffrage, Congress left that decision up to the Southern states. It also left Northern states free to continue the disfranchisement of Negroes, but it exempted them from the penalties inflicted on the Southern states for the same decision. The real concern of the franchise provisions of the Fourteenth Amendment was not with justice to the Negro but with justice to the North. The rebel states stood to gain some twelve seats in the House if all Negroes were counted as a basis of representation and to have about eighteen fewer seats if none were counted. The Amendment fixed apportionment of representation according to enfranchisement.

There was a great deal of justice and sound wisdom in the Fourteenth Amendment, and not only in the first section conferring citizenship on the Negro and protecting his rights, but in the other three sections as well. No sensible person could contend that the rebel states should be rewarded and the loyal states penalized in apportionment of representation by the abolition of slavery and the counting of voteless freedmen. That simply made no sense. Nor were there many, in the North at least, who could object to the temporary disqualification for office and ballot of such Southern officeholders of the old regime as were described in the third section. The fourth section asserting the validity of the national debt and avoiding the Confederate debts was obviously necessary. As it turned out these were the best terms the South could expect — far better than they eventually got — and the South would have been wise to have accepted them.

The tragic failure in statesmanship of the Fourteenth Amendment lay not in its terms but in the equivocal and pusillanimous way it was presented. Had it been made a firm and clear condition for readmission of the rebel states, a lot of anguish would have been spared that generation as well as later ones, including our own. Instead, in equivocal deference to states rights, the South was requested to approve instead of being compelled to accept. In this I think the moderates were wrong and Thaddeus Stevens was right. As W. R. Brock put it, "The onus of decision was passed to the Southern states at a moment when they were still able to defy Congress but hardly capable of taking a statesmanlike view of the future." [13] It was also the fateful moment when President Johnson declared war on Congress and advised the South to reject the Amendment. Under the circumstances, it was inevitable that the South should reject it, and it did so with stunning unanimity. Only thirty-two votes were cast for ratification in all the Southern legislatures. This spelled the end of any hope for the moderate position in the Republican leadership.

After two years of stalling and fumbling, of endless committee work and compromise, the First Reconstruction Act was finally adopted in the eleventh hour of the expiring Thirty-ninth Congress. Only after this momentous bill was passed, was it realized that it had been drastically changed at the last moment by amendments that had not been referred to or considered by committees and that had been adopted without debate in the House and virtually without debate in the Senate. In a panicky spirit of urgency, men who were ordinarily clear-headed yielded their better judgment to the demand for anything-better-than-nothing. Few of them liked what they got, and fewer still understood the implications and the meaning of what they had done. Even John Sherman, who gave his name to the bill, was so badly confused and misled on its effect that he underestimated by some 90 percent the number who would be disqualified from office and disfranchised. And this was one of the key provisions of the bill. It was, on the whole, a sorry performance and was far from doing justice to the intelligence and statesmanship and responsibility of the men who shaped and passed the measure.

One thing was at least clear, despite the charges of the South-

[13] W. R. Brock, *An American Crisis: Congress and Reconstruction, 1865–1867* (London, 1963), 149.

ern enemies and the Northern friends of the act to the contrary. It was not primarily devised for the protection of Negro rights and the provision of Negro equality. Its primary purpose, however awkwardly and poorly implemented, was to put the Southern states under the control of men loyal to the Union or men the Republicans thought they could trust to control those states for their purposes. As far as the Negro's future was concerned, the votes of the Congress that adopted the Reconstruction Act speak for themselves. Those votes had turned down Stevens' proposal to assure an economic foundation for Negro equality and Sumner's resolutions to give the Negro equal opportunity in schools, in homesteads, and full civil rights. As for the Negro franchise, its provisions, like those for civil rights, were limited. The Negro franchise was devised for the passage of the Fourteenth Amendment and setting up the new Southern state constitutions. But disfranchisement by educational and property qualifications was left an available option, and escape from the whole scheme was left open by permitting the choice of military rule. No guarantee of proportional representation for the Negro population was contemplated, and no assurance was provided for Negro officeholding.[14]

A sudden shift from defiance to acquiescence took place in the South with the passage of the Reconstruction Act of March 2, 1867. How deep the change ran it would be hard to say. The evidence of it comes largely from public pronouncements of the press and conservative leaders, and on the negative side from the silence of the voices of defiance. The mood of submission and acquiescence was experimental, tentative, and precarious at best. It can not be said to have predominated longer than seven months, from spring to autumn of 1867. That brief period was crucial for the future of the South and the Negro in the long agony of Reconstruction.

Southerners watched intently the forthcoming state elections

[14] G. Selden Henry, "Radical Republican Policy Toward the Negro During Reconstruction, 1862–1872" (Ph.D. dissertation, Yale, 1963), 204–217.

in the North in October. They were expected to reflect Northern reactions to Radical Reconstruction and especially to the issue of Negro suffrage. There was much earnest speculation in the South. "It may be," said the Charleston *Mercury,* "that Congress but represents the feelings of its constituents, that it is but the moderate mouthpiece of incensed Northern opinion. It may be that measures harsher than any . . . that confiscation, incarceration, banishment may brood over us in turn! But all these things will not change our earnest belief — that *there will be a revulsion of popular feeling in the North.*" [15]

Hopes were aroused first by the elections in Connecticut on April 1, less than a month after the passage of the Reconstruction Act. The Democrats won in almost all quarters. The radical *Independent* taunted the North for hypocrisy. "Republicans in all the great states, North and West, are in a false position on this question," it said. "In Congress they are for impartial suffrage; at home they are against it." In only six states outside the South were Negroes permitted to vote, and in none with appreciable Negro population. The *Independent* thought that "it ought to bring a blush to every white cheek in the loyal North to reflect that the political equality of American citizens is likely to be sooner achieved in Mississippi than in Illinois — sooner on the plantation of Jefferson Davis than around the grave of Abraham Lincoln!" [16] Election returns in October seemed to confirm this. Republican majorities were reduced throughout the North. In the New England states and in Nebraska and Iowa, they were sharply reduced, and in New York, New Jersey, and Maryland, the party of Reconstruction went down to defeat. Democrats scored striking victories in Pennsylvania and Ohio. In Ohio, Republicans narrowly elected the Governor by 8,000 votes but overwhelmed a Negro suffrage amendment by 40,000. In every state where the voters expressed themselves on the Negro suffrage issue, they turned it down.

Horace Greeley read the returns bluntly, saying that "the Ne-

[15] Charleston *Mercury,* quoted in *DeBow's Review,* XXXVI (September 1867), 250. [16] *Independent,* April 4, 18, 1867.

gro question lies at the bottom of our reverses. . . . Thousands have turned against us because we purpose to enfranchise the Blacks. . . . We have lost votes in the Free States by daring to be just to the Negro." [17] The *Independent* was quite as frank. "Negro suffrage, as a political issue," it admitted, "never before was put so squarely to certain portions of the Northern people as during the late campaigns. The result shows that the Negro is still an unpopular man." [18] Jay Cooke, the conservative financier, wrote John Sherman that he "felt a sort of intuition of coming disaster — probably growing out of a consciousness that other people would feel just as I did — disgust and mortification at the vagaries into which extremists in the Republican ranks were leading the party." [19]

To the South, the Northern elections seemed a confirmation of their hopes and suspicions. The old voices of defiance and resistance, silent or subdued since March, were lifted again. They had been right all along, they said. Congress did not speak the true sentiment of the North on the Negro and Reconstruction. President Johnson had been the true prophet. The correct strategy was not to seek the Negro vote but to suppress it, not to comply with the Reconstruction Acts but to subvert them. The New York *Times* thought that "the Southern people seem to have become quite beside themselves in consequence of the *quasi* Democratic victories" in the North, and that there was "neither sense nor sanity in their exultations." [20] Moderates such as Governor James W. Throckmorton of Texas, who declared he "had advocated publicly and privately a compliance with the Sherman Reconstruction Bill," were now "determined to defeat" compliance and to leave "no stone unturned" in their efforts.[21]

The standard Southern reply to Northern demands was the endlessly reiterated charge of hypocrisy. Northern radicals, as a Memphis conservative put it, were "seeking to fasten what they

17 Quoted in *ibid.*, Nov. 21, 1867. 18 *Ibid.*, Nov. 14, 1867.

19 Jay Cooke to John Sherman, Oct. 12, 1867, John Sherman Papers #28298, Library of Congress. 20 New York *Times*, Oct. 19, 1867.

21 J. W. Throckmorton to B. H. Epperson, Dec. 19, 1867, Epperson Papers, University of Texas Archives.

themselves repudiate with loathing upon the unfortunate people of the South." And he pointed to the succession of Northern states that had voted on and defeated Negro suffrage.[22] A Raleigh editor ridiculed Republicans of the Pennsylvania legislature who voted 29 to 13 against the franchise for Negroes. "This is a direct confession, by Northern Radicals," he added, "that they refuse to grant in Pennsylvania the '*justice*' they would enforce on the South. . . . And this is Radical meanness and hypocrisy — this their love for the negro." [23]

There was little in the Republican presidential campaign of 1868 to confute the Southern charge of hypocrisy and much to support it. The Chicago Platform of May on which General Grant was nominated contained as its second section this formulation of the double standard of racial morality: "The guaranty by Congress of equal suffrage to all loyal men at the South was demanded by every consideration of public safety, of gratitude, and of justice, and must be maintained; while the question of suffrage in all the loyal [i.e., Northern] States properly belongs to the people of those States." Thus Negro *dis*franchisement was assured in the North along with enfranchisement in the South. No direct mention of the Negro was made in the entire platform, and no mention of schools or homesteads for freedmen. Neither Grant nor his running-mate Schuyler Colfax was known for any personal commitment to Negro rights, and Republican campaign speeches in the North generally avoided the issue of Negro suffrage.

Congress acted to readmit seven of the reconstructed states to the Union in time for them to vote in the presidential election and contribute to the Republican majority. In attaching conditions to readmission, however, Congress deliberately refrained from specifying state laws protecting Negroes against discrimination in jury duty, officeholding, education, intermarriage, and a wide range of political and civil rights. By a vote of 30 to 5, the Senate defeated a bill attaching to the admission of Arkansas

[22] Memphis *Avalanche,* Nov. 10, 1867.
[23] Raleigh *Daily Sentinel,* March 11, 1868.

the condition that "no person on account of race or color shall be excluded from the benefits of education, or be deprived of an equal share of the moneys or other funds created or used by public authority to promote education. . . ." [24]

Not until the election of 1868 was safely behind them did the Republicans come forward with proposals of national action on Negro suffrage that was to result in the Fifteenth Amendment. They were extremely sensitive to Northern opposition to enfranchisement. By 1869, only seven Northern states had voluntarily acted to permit the Negro to vote, and no state with a substantial Negro population outside the South had done so. Except for Minnesota and Iowa, which had only a handful of Negroes, every postwar referendum on the subject had gone down to defeat.

As a consequence moderates and conservatives among Republicans took over and dominated the framing of the Fifteenth Amendment and very strongly left their imprint on the measure. Even the incorrigibly radical Wendell Phillips yielded to their sway. Addressing other radicals, he pleaded, ". . . for the first time in our lives we beseech them to be a little more *politicians* and a little less *reformers*." The issue lay between the moderates and the radicals. The former wanted a limited, negative amendment that would not confer suffrage on the freedmen, would not guarantee the franchise and take positive steps to protect it, but would merely prohibit its denial on the grounds of race and previous condition. Opposed to this narrow objective were the radicals who demanded positive and firm guarantees, federal protection, and national control of suffrage. They would take away state control, North as well as South. They fully anticipated and warned of all the elaborate devices that states might resort to — and eventually did resort to — in order to disfranchise the Negro without violating the proposed amendment. These included such methods — later made famous — as the literacy and property tests, the understanding clause, the poll tax, as well as elaborate and difficult registration tricks and

[24] Edward McPherson (ed.), *The Political History of the United States . . . During . . . Reconstruction* (Washington, 1871), 337–41.

handicaps. But safeguards against them were all rejected by the moderates. Only four votes could be mustered for a bill to guarantee equal suffrage to all states, North as well as South. "This amendment," said its moderate proponent Oliver P. Morton, "leaves the whole power in the State as it exists, now, except that colored men, shall not be disfranchised for the three reasons of race, color, or previous condition of slavery." And he added significantly, "They may, perhaps, require property or educational tests." [25] Such tests were already in existence in Massachusetts and other Northern states, and the debate made it perfectly apparent what might be expected to happen later in the South.

It was little wonder that Southern Republicans, already faced with aggression against Negro voters and terribly apprehensive about the future, were intensely disappointed and unhappy about the shape the debate was taking. One of their keenest disappointments was the rejection of a clause prohibiting denial or abridgment of the right of officeholding on the ground of race. It is also not surprising that Southern white conservatives, in view of these developments, were on the whole fairly relaxed about the proposed Fifteenth Amendment. The shrewder of them, in fact, began to realize that the whole thing was concerned mainly, not with the reconstruction of the South, but with maneuvers of internal politics in the Northern states. After all, the Negroes were already fully enfranchised and voting regularly and solidly in all the Southern states, their suffrage built into state constitutions and a condition of readmission to the Union.

Were there other motives behind the Fifteenth Amendment? The evidence is somewhat inferential, but a recent study has drawn attention to the significance of the closely divided vote in such states as Indiana, Ohio, Connecticut, New York, and Pennsylvania. The Negro population of these states was small, of course, but so closely was the white electorate in them divided between the two major parties that a small Negro vote could

[25] Quoted in Henry, "Radical Republican Policy Toward the Negro," 255.

often make the difference between victory and defeat. It was assumed, of course, that this potential Negro vote would be reliably Republican. Enfranchisement by state action had been defeated in all those states, and federal action seemed the only way. There is no doubt that there was some idealistic support for Negro enfranchisement, especially among antislavery people in the North. But it was not the antislavery idealists who shaped the Fifteenth Amendment and guided it through Congress. The effective leaders of legislative action were moderates with practical political considerations in mind — particularly that thin margin of difference in partisan voting strength in certain Northern states. They had their way, and they relentlessly voted down all measures of the sort the idealists, such as Senator Sumner, were demanding.[26]

For successful adoption the amendment required ratification by twenty-eight states. Ratification would therefore have been impossible without support of the Southern states, and an essential part of that had to come by requiring ratification as a condition of readmission of Virginia, and perhaps of Mississippi and Georgia as well.[27]

The Fifteenth Amendment has often been read as evidence of renewed notice to the South of the North's firmness of purpose, as proof of its determination not to be cheated of its idealistic war aims, as a solemn rededication to those aims. Read more carefully, however, the Fifteenth Amendment reveals more deviousness than clarity of purpose, more partisan needs than idealistic aims, more timidity than boldness.

Signals of faltering purpose in the North, such as the Fifteenth Amendment and state elections in 1867, were not lost on the South. They were assessed carefully and weighed for their implications for the strategy of resistance. The movement of counterreconstruction was already well under way by the time the amendment was ratified in March 1870, and in that year, the reactionary movement took on new life in several quarters. Fun-

[26] William Gillette, *The Right to Vote: Politics and the Passage of the Fifteenth Amendment* (Baltimore, 1965), *passim*. [27] *Ibid.*, 92.

damentally it was a terroristic campaign of underground organizations, the Ku Klux Klan and several similar ones, for the intimidation of Republican voters and officials, the overthrow of their power, and the destruction of their organization. Terrorists used violence of all kinds, including murder by mob, by drowning, by torch; they whipped, they tortured, they maimed, they mutilated. It became perfectly clear that federal intervention of a determined sort was the only means of suppressing the movement and protecting the freedmen in their civil and political rights.

To meet this situation, Congress passed the Enforcement Act of May 30, 1870, and followed it with the Second Enforcement Act and the Ku Klux Klan Act of 1871. These acts on the face of it would seem to have provided full and adequate machinery for the enforcement of the Fifteenth Amendment and the protection of the Negro and white Republican voters. They authorized the President to call out the army and navy and suspend the writ of habeas corpus; they empowered federal troops to implement court orders; and they reserved the federal courts' exclusive jurisdiction in all suffrage cases. The enforcement acts have gone down in history with the stereotypes "infamous" and "tyrannical" tagged to them. As a matter of fact, they were consistent with tradition and with democratic principle. Surviving remnants of them were invoked in recent years to authorize federal intervention at Little Rock and at Oxford, Mississippi. They are echoed in the Civil Rights Acts of 1957 and 1960, and they are greatly surpassed in the powers conferred by the Civil Rights Act of 1964 and the Voting Rights Act of 1965.

Surely this impressive display of federal power and determination, backed by gleaming steel and judicial majesty, might be assumed to have been enough to bring the South to its senses and dispel forever the fantasies of Southern intransigents. And in fact, historians have in the main endorsed the assumption that the power of the Klan was broken by the impact of the so-called Force Bills.

The truth is that, while the Klan was nominally dissolved, the campaign of violence, terror, and intimidation went forward virtually unabated, save temporarily in places where federal power was displayed and so long as it was sustained. For all the efforts of the Department of Justice, the deterioration of the freedman's status and the curtailment and denial of his suffrage continued steadily and rapidly. Federal enforcement officials met with impediments of all sorts. A close study of their efforts reveals that "in virtually every Southern state . . . federal deputy marshals, supervisors of elections, or soldiers were arrested by local law-enforcement officers on charges ranging from false arrest or assault and battery to murder." [28]

The obvious course for the avoidance of local passions was to remove cases to federal courts for trial, as provided under a section of the First Enforcement Act. But in practice this turned out to be "exceedingly difficult." And the effort to find juries that would convict proved often to be all but impossible, however carefully they were chosen, and in whatever admixture of color composed them. The most overwhelming evidence of guilt proved unavailing at times. Key witnesses under intimidation simply refused to testify, and those that did were known to meet with terrible reprisals. The law authorized the organization of the *posse comitatus* and the use of troops to protect juries and witnesses. But in practice the local recruits were reluctant or unreliable, and federal troops were few and remote and slow to come, and the request for them was wrapped in endless red tape and bureaucratic frustration.[29]

All these impediments to justice might have been overcome had sufficient money been made available by Congress. And right at this crucial point, once again, the Northern will and purpose flagged and failed the cause they professed to sustain. It is quite clear where the blame lies. Under the new laws, the cost of maintaining courts in the most affected districts of the South soared tremendously, quadrupled in some. Yet Congress starved

[28] Everette Swinney, "Enforcing the Fifteenth Amendment, 1870–1877," *Journal of Southern History*, XXVIII (May 1962), 210.
[29] *Ibid.*, 210–11.

the courts from the start, providing only about a million dollars a year — far less than was required. The Attorney General had to cut corners, urge economy, and in 1873 instruct district attorneys to prosecute no case "unless the public interest imperatively demands it." An antiquated judicial structure proved wholly inadequate to handle the extra burden and clear their dockets. "If it takes a court over one month to try five offenders," asked the Attorney General concerning 420 indictments in South Carolina, "how long will it take to try four hundred, already indicted, and many hundreds more who deserve to be indicted?" He thought it "obvious that the attempt to bring to justice even a small portion of the guilty in that state must fail" under the circumstances. Quite apart from the inadequacy and inefficiency of the judicial structure, it is of significance that a majority of the Department of Justice officers in the South at this time, despite the carpetbagger infusion, were Southern-born. A study by Everette Swinney concludes that "some marshals and district attorneys were either sensitive to Southern public opinion or in substantial agreement with it." The same has been found true of numbers of federal troops and their officers on duty in the South.[30] Then in 1874 an emasculating opinion of the Supreme Court by Justice Joseph P. Bradley in *United States* v. *Cruikshank et al.* cast so much doubt on the constitutionality of the enforcement acts as to render successful prosecutions virtually impossible.

There is also sufficient evidence in existence to raise a question about how much the Enforcement Acts were intended all along for application in the policing of elections in the South, as against their possible application in other quarters of the Union. As it turned out, nearly half of the cost of policing was applied to elections of New York City, where Democratic bosses gave the opposition much trouble. Actually the bulk of federal expenditures under the Enforcement Acts was made in the North which leads one student to conclude that their primary object from the start was not the distraught South under reconstruc-

[30] *Ibid.*, 212–16.

tion, but the urban strongholds of the Democrats in the North.[31] Once again, as in the purposes behind the Fifteenth Amendment, one is left to wonder how much Radical Reconstruction was really concerned with the South and how much with the party needs of the Republicans in the North.

Finally, to take a longer view, it is only fair to allow that if ambiguous and partisan motives in the writing and enforcing of Reconstruction laws proved to be the seeds of failure in American race policy for the earlier generations, those same laws and constitutional amendments eventually acquired a wholly different significance for the race policy of a later generation. The laws outlasted the ambiguities of their origins. While the logic that excuses and vindicates the failures of one generation by reference to the successes of the next has always left something to be desired. It is, nevertheless, impossible to account fully for such limited successes as the Second Reconstruction can claim without acknowledging its profound indebtedness to the First.

[31] Robert A. Horn, "National Control of Congressional Elections" (Ph.D. dissertation, Princeton, 1942), 143, 154–55, 183–87.

7

A Southern Brief for Racial Equality

It is now more than three quarters of a century since Lewis Harvie Blair of Richmond, Virginia, published his book in defense of Negro equality. An uncompromising attack on racial segregation, discrimination, and injustice of any kind, his book demanded full civil rights for the Negro, equal access to hotels, theaters, and all places of amusement and public accommodations, unrestricted franchise and political rights, as well as integration of churches and public schools. Accompanying these demands was a blistering and iconoclastic attack on the dogmas of white supremacy and Negro inferiority, the plantation legend of slavery, the paternalistic tradition of race relations, the black-domination picture of Reconstruction, and the complacent optimism of the New South school of economists.

Published in 1889, Blair's *Prosperity of the South Dependent upon the Elevation of the Negro* attracted little attention and had little influence. It was soon forgotten and has been neglected ever since. The quick oblivion is not hard to explain. The book appeared not long before the great racist reaction that overtook the country in the 1890's. In the South, the reaction found expression in white supremacy propaganda, segregation laws, poll taxes, literacy and property tests, white primaries and other devices for disfranchising the Negroes. It was accompanied by an increase in lynching, riots, and other forms of violence against the minority race. It resulted in driving the Negro from all forums and avenues of political life, in stripping him of many of the civil rights and defenses he had gained through the Reconstruction amendments, and in reducing him to a despised and segregated outcast.

By the end of the century, the South had reached a consensus on race policy. Its mind was closed. The debate was over. Dissent was frowned down or smashed. Conformity was demanded of all. The ensuing rigidity of regional attitude is reminiscent of that which occurred over the slavery issue in the early 1830's. And under these circumstances, a book by a Southerner of authentic lineage and high standing that challenged every dogma of the new consensus from top to bottom was about as welcome and popular as a red-hot abolitionist tract by a comparable Southern author in the 1830's.

But in 1889, the great freeze had not yet taken place. Alternatives were still available. Real choices had to be made. Many of those issues about which there was soon to be such stiff conformity and incorrigible rigidity were still open questions. And in Virginia, there was rather more hesitancy about closing off the debate and clamping down conformity than there was in other Southern states.

Charles E. Wynes, the most recent and thorough student of race relations in Virginia, finds that "the most distinguishing factor in the complexity of social relations between the races was that of inconsistency. From 1870–1900, there was no generally accepted code of racial mores." During that period of thirty years, according to this scholar, "at no time was it the general demand of the white populace that the Negro be disfranchised and white supremacy made the law of the land." The era of Jim Crow was still to come, and its first formal appearance in Virginia was not until 1900, when a law requiring the separation of the races on railroad cars was adopted. Up until that time, according to Mr. Wynes, "the Negro sat where he pleased and among the white passengers on perhaps a majority of the state's railroads." The same was true of streetcars and, with greater variation and more exceptions, of other public accommodations and places of entertainment. While he often encountered rebuff and even eviction, "occasionally the Negro met no segregation when he entered restaurants, bars, waiting rooms, theatres, and other public places of amusement." Whatever the risks and uncertainties and the crosscurrents of ambivalence and ambiguity, there was still a considerable range of flexibility, tolerance, and uninhibited contact and association in relations between the races in Virginia.[1]

In the political life of the state during this period, the Negroes, being a majority of the population in forty of the ninety-nine counties, played a prominent and sometimes crucial part. They held numerous public offices, elective and appointive. There were Negro members of the General Assembly, the oldest representative legislative body in the New World, in every session from 1869 to 1891.[2] Negroes voted in large numbers, and while they were overwhelmingly Republican, their votes were sought by both of the major parties and with striking success by two important third parties in the period, the Readjusters and the Populists. A split between radicals and moderates in the Republican party deprived the Negroes of the prominent role they played during Reconstruction in some other Southern states. But they compensated for this in large measure by the part they played in the Readjuster party, which drove the conservatives from power in 1879 and took control of the state until 1883. Combining with impoverished and discontented white farmers and workers in support of the Readjusters, the Negroes assisted in giving Virginia a foretaste of Populism and the most liberal reform administration the state ever had — before or since. Among the relics of the old regime swept away by the reformers were the whipping post, the poll tax as a voting requirement, and official connivance at dueling. In addition, the Readjusters rejuvenated the impoverished public schools, founded a Negro col-

[1] Charles Wynes, *Race Relations in Virginia, 1870–1902* (Charlottesville, Va., 1961), especially 68, 149–150.

[2] Luther P. Jackson, *Negro Office-Holders in Virginia, 1865–1895* (Norfolk, Va., 1945).

lege, chartered labor unions, reformed the laws to tax privileged corporations, and in general reversed Virginia's retreat to antebellum reactionism and advanced the state to a leading position among Southern states.

Returning to power in 1884, the conservatives rushed through a new election law designed to defraud Negro voters and invite ballot-box corruption. The law was effectively used for this purpose, but Negro political strength still remained formidable. In the presidential election of 1888, the Democratic candidate won by a narrow margin, with 151,979 for Cleveland, and 150,449 for Harrison. In the same year the first (and last) Negro, John Mercer Langston, was elected to Congress from Virginia. At the same time the people of the state voted down a proposal to hold a convention to amend the liberal Reconstruction Constitution of 1868, a move that would have made possible the curtailment of Negro suffrage.

Such was the temper of mind, the balance of forces, and the accommodation of races in Virginia in 1889. It was one of the pivotal moments of history when public commitment and decision were still in suspense, when to all appearances the balance could swing either way. Voices of reaction, racism, and fanaticism were already calling for extreme measures — for disfranchisement, for segregation, for rigid conformity of white supremacy and a closed society. Their influence was already being felt in the lower South, and they were at work in the upper South as well. But in most of the Southern states, and in Virginia particularly, there was still strong resistance to the fanatics. There were even Southerners who denounced the existing compromise of racial accommodation and called for radical advance to a new order of equal rights and racial justice. They were few in number, but there was a willingness to hear them out. The situation was, for the moment, still fluid. There was a disposition to suspend judgment, to consider alternatives, and to ponder the future with an open mind.

Assuming the normal risks of historical comparisons, one might profitably think in this connection of another pivotal moment in racial relations that occurred in Virginia nearly two generations earlier, in 1832. The slave insurrection of August, 1831, in Southampton County led by Nat Turner had horrified and shaken Virginia profoundly, and the shock permeated the whole South. Disturbed and sobered, Virginians opened a searching, unrestrained, and thorough public debate on the nature and evils of slavery and the needs and means of abolishing the institution. When the legislature met in Richmond in December a committee was appointed to consider the question, and in January, a formal debate explored every aspect of slavery — its incompatibility with natural rights and human dignity, its responsibility for economic backwardness, its damage to white manners and morals as well as to Negro welfare and personality. "The institution was denounced as never before," according to one historian; "it was condemned wholesale fashion by legal representatives of a slave-holding people." Among the opponents of slavery were the governor of the state, the editors of the two leading papers of Richmond, a son of John Marshall, and a grandson of Thomas Jefferson. It has been called the "final and most brilliant of the Southern attempts to abolish slavery," but it failed. A majority of the legislature favored some form of abolition, but was unable to agree on the time or the method. The moment passed. Virginia took the path that led to a closed society, a rigid conformity, and a last-ditch defense of the peculiar institution.[3]

Two generations later, in the new debates over the future of race relations, Lewis Blair occupied an extreme position toward the left. But he was not without Southern support for many of his views. Writing in 1886, Noah K. Davis, a philosophy professor at the University of Virginia, said: "What shall we care whether the laws, so they be laws, be made by white or black. We want in Congress men of capacity, honesty, strength. Color is nonessential. . . . The time may come when a Negro shall be our Secretary of State; and who will be foolish enough to

[3] Joseph C. Roberts, *The Road from Monticello: A Study of the Virginia Slavery Debate of 1832* (Durham, N.C., 1941), 11–18.

object?" He drew the line at social intercourse, but predicted that, "We shall see, or our children shall see, white servants and laborers under Negro employers. Resistance would be vain, and regret senseless. Brains, not color, must settle rank." [4]

Of greater importance as a Southern champion of Negro rights was the Louisiana novelist George W. Cable. Writing and lecturing with great conviction and passion through the 1880's and early 1890's, Cable spoke out against every form of injustice toward the Negro — whether by police or church, in schools or in prisons, at the polls or on the job, by mobs or by courts. If he spared Southern sensibilities on the issue of social intercourse, it was out of strategy.[5]

A more negative and phlegmatic restraint upon the advance of racist fanaticism came from some of Blair's own social class. They shared none of his equalitarian convictions, but from the security of their own social position they could not regard the Negro as a threat to the social order and could not understand the racial phobias and hysterias that afflicted lower classes of whites. They sometimes found the Negro vote adaptable to their political maneuvers. Out of their paternalistic tradition, they deplored violence against Negroes and occasionally provided effective protection against movements to disfranchise, humiliate, or segregate them. Even conservative papers such as the Charleston *News and Courier* could, as late as 1898, print indignant editorials on the stupidity of the demand for Jim Crow laws.

Unexpected support for Negro rights from a lower order of white society was recruited by the Populist movement in the early 1890's. The Populist approach to the Negro was somewhat paternalistic and humanitarian but more pragmatic and economic. Caught in the pinch of agrarian depression, white and black farmers formed a political brotherhood on the kinship of common grievances and common oppressors. In their efforts to

[4] Noah K. Davis, "The Negro in the South," *Forum* I (April 1886), 133–135.

[5] George W. Cable, *The Silent South* (New York, 1885) and *The Negro Question* (New York, 1890).

win and hold their Negro supporters, the Populists fought for Negro civil and political rights and offered the Negroes a warmer political fellowship and greater political equality than they had ever received before.

In none of these camps — the patricians, the Populists, the liberals, or the humanitarians — could there be found a champion of Negro rights who went quite so far in the 1880's and 1890's as the curious Richmond merchant aristocrat, Lewis Blair. Indeed, it would be difficult in that period to find his match in any part of the country.

Lewis Harvie Blair was born in Richmond on June 21, 1834, and died there on November 26, 1916. By the time his generation came along, the branches of the Blair family tree were laden with distinguished names and public honors. The Blair name began to figure prominently in colonial times and continued to grow in distinction through the Civil War years. It was borne by prominent theologians, college presidents, generals, editors, and politicians of national and local importance, including congressmen, senators, cabinet members, and presidential aspirants.

The branch of the family from which Lewis was descended was started in America by John Blair, his great-grandfather, who was born in Ireland in 1720 and emigrated as a child with his older brother Samuel.[6] Both brothers received a classical education at an academy near Philadelphia, both became licensed and ordained Presbyterian ministers at an early age, and both fell under the influence of the evangelist George Whitefield and the Great Awakening. John Blair held the first chair of theology at Princeton, then called the College of New Jersey, from which he had received an honorary degree in 1760. He also served Princeton as a trustee, as vice-president, and for two years as acting president.

[6] For this and much genealogical information that follows, I rely mainly on Louisa Coleman Gordon Blair (ed.), *Blairs of Richmond, Virginia: The Descendants of Reverend John Durburrow Blair and Mary Winston Blair, His Wife* (Richmond, 1933).

Two of John's sons moved to Virginia before the end of the Revolution. James left the state early to settle in Kentucky and found the powerful political family that included his son Francis Preston Blair, Francis Preston, Jr., and Montgomery. John Durburrow Blair, the other son, after graduating in 1775 from Princeton, where he served as tutor, and seeing service with the Revolutionary army, came to Hanover County, Virginia, in 1780 at the age of twenty-one as president of Washington Henry Academy. About ten years later, he moved to Richmond to open a classical school and serve the city as its only Presbyterian minister. Having no church in which to preach, he resorted to the Hall of the House of Delegates in Mr. Jefferson's newly completed State Capitol and for years met his congregation in that elegantly classical temple. By this time, the fires of the Great Awakening had burned low and appeared to have left few sparks in the bosom of the Reverend Mr. Blair. His disposition was described as "amiable, his temper sweet," and he is reported to have written "many merry notes in verse to his friends that did not lack wit." An affectionate memoir quotes scores of his letters written in rhyming jingles, some in Latin, that give an engaging picture of the hunting, fishing parson. His ministry did not go uncriticized. His critics called him "a wine bibber, a friend of publicans and sinners," and declared that "the fruits of his ministry were not apparent." [7]

The minister's son, John Geddes Blair, born in 1787, was the cashier of a Richmond bank and appears to have contented himself with simple pleasures and a rather active domestic life. He was the father of thirteen children, of whom Lewis Harvie was the eleventh. Lewis described his father as a "courteous gentleman," a man of "refined taste and good intelligence" with minor gifts in the arts. "He played sweetly on the flute," remembered Lewis, "sang fairly well, and wrote quite pleasing lines. He was sociable, belonged to an instrumental musical society which met at his house, to a quoit club, composed of prominent gentlemen,

[7] *Ibid.*, 87–96; see also George Wythe Munford, *The Two Parsons* (Richmond, 1884), *passim.*

and to the Richmond Blues, with whom he took the field in the War of 1812." One enigmatic sidelight on the character of his father is furnished in a description Lewis wrote of his mother, Sarah Anne Eyre Heron Blair. He describes her as being "pretty as a maiden, matron, and old lady," and then adds: "She was a firmer texture than father." [8]

John, the flute player of infirm texture, departed this life on March 7, 1851, leaving a widow and ten surviving children to be supported by an estate that was somewhat less than adequate to the demands upon it. As a result, Lewis, not quite seventeen at the time of his father's death, dropped out of school and went to work, never to continue his education. During the ten years preceding the Civil War he turned his hand to a number of jobs under private and government employment. His first was that of clerk to his brother, a captain in the army, then stationed on the Texas frontier, where Lewis served from 1851 to 1855. He then returned to Richmond to work for a merchant a couple of years and in 1857 took a government job as assistant to a lighthouse engineer on the Great Lakes.[9]

Lewis quit the lonely lighthouse work early in 1860 and went back to Richmond seeking to establish himself in business. Before he had got a start, however, the war came and disarranged his life again. He enlisted as a private in the Confederate army in March 1862, just ahead of the draft. He served undramatically with artillery and cavalry units for two years and for the remainder of the war as adjutant to the commander of a cavalry division. What the experience meant to him at the time is unclear, but his later description of his military service is revealing: "more than three years wasted in the vain effort to maintain that most monstrous institution, African slavery, the real, tho' States Rights were the ostensible cause of the War." [10]

[8] "Autobiography of Lewis H. Blair," quoted in Blair (ed.), *Blairs of Richmond*, 10n.

[9] Blair's recollection of the Texas adventure is preserved by Charles E. Wynes (ed.), "Lewis Harvie Blair: Texas Travels, 1851–1855," *Southwestern Historical Quarterly* LXVI (1962), 262–270.

[10] A full account of Blair's war career, including his quoted comment

After Appomattox, he returned to find the city of his fathers in ashes and his state and region in ruins. He was nearly thirty-one that blighted spring of 1865, with a start in life still to make and a prospect around him bleak enough to discourage even the stout of heart. But Lewis Blair had the sanguine disposition of the tribe of enterprise and business, and instead of despair, he felt only a sense of release and optimism. In the opening passage of his book on the Negro, he described his hopes on returning to the ruins of Richmond. "I fondly imagined a great era of prosperity for the South," he wrote. "Guided by history and by a knowledge of our people and our climatic and physical advantages, I saw in anticipation all her tribulations ended, all her scars healed, and all the ravages of war forgotten, and I beheld the South greater, richer and mightier than when she molded the political policy of the whole country."

Blair admitted twenty-four years after the war that his hopes had been cruelly disappointed. He wrote then that, "instead of beholding the glorious South of my imagination, I see her sons poorer than when war ceased his ravages, weaker than when rehabilitated with her original rights, and with the bitter memories of the past smoldering, if not rankling, in the bosoms of many." He had his own explanation for the disappointment of his hopes and the failure of the South, but that is a later story. The point is that it was during the bitter, iron years of that quarter of a century, for the most part years of depression, poverty, and economic stagnation in the South, that Blair struggled to get his start, founded his fortune, and regained the position his family had once maintained in Richmond. The means he used were several, but the first was a retail, and then a wholesale, grocery business. Then came the manufacture and sale of shoes with headquarters in Richmond and outlets in various towns, and later a thriving business in real estate.

Judging from the style in which he subsequently lived, Blair's

on it, is in Charles E. Wynes, "Lewis Harvie Blair, Virginia Reformer: The Uplift of the Negro and Southern Prosperity," *Virginia Magazine of History and Biography,* LXXII (January 1964), 3–18.

struggles did not go unrewarded. His handsome house at 511 East Grace Street was an old one built by Corbin Warwick, who imported its elaborate white Italian marble mantels. Blair added a fashionable mansard-style top story and "a rather modern front porch with a tiled floor." In the eighties and nineties Grace Street was lined with the homes of "old families" and Richmond's great and near-great. Next door to the "Warwick-Blair" house, as it was called, was the home of George E. Barksdale, and nearby lived Randolphs, Branches, Cabells, Pages, Lees, Carys, and "from time to time scores of prominent families." [11] Abner Harvey, a partner in one of Blair's several business enterprises, occupied "one of the largest and handsomest houses," and several of Blair's numerous brothers and sisters were suitably established in the city. Blair married twice. His first wife, Alice Wayles Harrison, bore him seven children; his second, Mattie Ruffin Feild, produced four.

It is clear that the Blairs carried their heads pretty high in Richmond. Lewis more than recouped the family fortunes and status, for it is doubtful that any of the three preceding generations of American Blairs lived in finer style or occupied higher social positions in the city. Whatever may explain the rebellion of this Richmond aristocrat, it was hardly loss of status. Far from being an alienated outcast or a frustrated climber, Lewis Blair was a man of wealth and power and position. More than that, he was at one and the same time "self-made" and "old family" in a society and time when both distinctions counted — especially the latter. And Blair's personality and appearance emphasized identification with the latter. A kinsman who did not share his radical views nevertheless described him as "a man of distinguished appearance and of a courtesy in manner that seemed to belong to an even older time." [12]

For all his lack of formal education, Blair was not without

[11] Robert Beverly Munford, Jr., *Richmond Homes and Memories* (Richmond, 1936), 194–201. Sometime later Blair moved to a house on Monument Avenue. [12] *Ibid.*, 198.

cultivated tastes in arts and letters. The same kinsman who conceded his courtly manners wrote: "He was also a gentleman of cultivated intellect who appreciated works of art, and in his Grace Street home were a good many paintings of merit." There were evidently books of merit in the establishment on Grace Street as well, for its master was a reader of some scope and evidently liked to think of himself as abreast of the times. There is little in what he wrote to suggest that he cultivated the taste of his forebears for the classics. He was more interested in modern literature on philosophy, politics, and economics — what the old school called "moral philosophy." He founded his beliefs, he said, on the "school of Adam Smith, Herbert Spencer, and [Ernst] Haeckel." He was something of a skeptic and was reported to entertain "very decided opinions in religious matters, opinions which differ decidedly from those of the majority of his friends and neighbors." [13]

In political matters, with the important exception of questions of civil rights and racial policy, Blair seems to have been more in accord with his neighbors. Enough so at least to have been elected treasurer of the Democratic Committee of Richmond, though his stand for desegregation of the public schools brought a public demand for his resignation from that office in 1887.[14] The year before, however, G. P. Putnam's Sons published his first book, *Unwise Laws,*[15] a free-trade tract that advanced views familiar enough to the antiprotectionist wing of the Democratic party in the South. He took the extreme laissez-faire position against protective tariff of any kind whatever and held that "all taxes levied shall be exclusively for the benefit of the government and none for the benefit of industries whether infant or ancient." Acknowledging that such a reform would require dras-

[13] Lyon G. Tyler (ed.), *Encyclopedia of Virginia Biography* (New York, 1915), III, 185. [14] Wynes, *Race Relations in Virginia,* 126.

[15] Lewis H. Blair, *Unwise Laws: A Consideration of the Operations of a Protective Tariff Upon Industry, Commerce, and Society* (New York, 1886).

tic adjustments in the economy and drive some favored industries out of business, he justified his views with arguments drawn from classical economists of the day.

As a polemicist, Blair liked to assume the tone of a hard-boiled realist, a materialist who relied little on the nobler impulses of his fellow man and addressed himself to motives of self-interest. In his opinion "all experience proves that human nature must be interested on its material side before its intellectual and moral interests can be permanently stimulated." [16] The title of his book on the Negro as well as the arguments he stressed illustrate his theory. "You will observe," he pointed out, "that I have appealed little, if at all, to right, justice, morals, or religion, but that the burden of the argument has been dollars and cents." Moreover, "this key-note of appeal to the pocket . . . was deliberately struck." He called for drastic moral reform not "because this was great or noble, but simply because it would be profitable." [17] There was nevertheless a good deal of the iron glove on the velvet hand in all this. For behind the rather ostentatious "hard doctrine" he preached, it is not difficult to detect a fierce moral indignation over injustice and strong impulses of ethical idealism.

The Richmond reformer had every reason to be aware of the unpopularity of his racial doctrines in the South. As one of his Virginia contemporaries mildly remarked, "Mr. Blair utters views that in all probability will never achieve popularity south of the Potomac." Blair knew he would be accused of outrageous motives but said he would "patiently bear the odium attached to such charges." "It may seem both invidious and presumptuous to attempt this role," he wrote, "but while frankly admitting that I may be presumptuous, my Southern ancestry, birth, rearing,

[16] Tyler (ed.), *Encyclopedia of Virginia Biography*, III, 185.

[17] These and all subsequent quotations in this chapter, unless otherwise noted, are from my edition of Blair's book (see note 15) under the title, *A Southern Prophecy: The Prosperity of the South Dependent Upon the Elevation of the Negro* (Boston, 1964).

residence and interest preserve me from the charge of invidiousness." At least no one could call him an outsider.

The role of the iconoclast was obviously congenial to the Blair temperament. This was evident from the zest with which in his book on the Negro problem[18] he pitched into his attack on cherished and hallowed Southern myths, prejudices, credos, anything that stood in the way of the new social doctrine he preached.

His first target was the brightest and most conspicuous on the contemporary scene — the New South gospel. In 1889, this doctrine had reached a peak of popularity on the eve of the death of its major prophet, Henry W. Grady of Atlanta, the famous orator and editor. Grady left scores of ardent apostles and a host of converts. A message of such cheerful optimism held great attraction for a hope-starved people. Its adherents proclaimed that the cloud of depression and poverty was lifting, that prosperity was rolling in, that great cities like those of the golden East were springing up out of ashes, and manufacturing industries were growing at a magic pace. The great leap forward into the age of industrialization could be accomplished without any painfully protracted period of capital accumulation merely by will power and enough publicity to attract eager Northern and foreign investment capital to abundant natural resources and the cheap labor supply. The era of a depressed, underdeveloped, colonial, raw-material economy was at an end. The states of the late Confederacy, moreover, could march triumphantly into the promised land of industrialization without abandoning many of the values and loyalties and habits of the past. These included white supremacy, the comfortable assurance that the degraded mass of Negroes knew their place and would keep to it, the inexhaustible patience of the impoverished whites, and a large assortment of sentimental ties to the Lost Cause and the old regime.

[18] The book was foreshadowed by a series of articles Blair wrote for the New York *Independent* in the summer of 1887.

Blair struck contemptuously at the "brag, and strut, and bluster" of the boosters and the prosperity propaganda of such journals as the Baltimore *Manufacturer's Record.* "Judging by these sheets," he wrote, "one would naturally imagine that the South is a region where poverty is unknown and where everybody is industriously and successfully laying up wealth," that "the South is indeed a happy Arcadia." It was true that New South propaganda had managed to persuade a few Northern journalists and capitalists that it was indeed a "happy Arcadia." But "these gentlemen, having been hurried through hundreds of miles in luxurious palace coaches, have practically been blindfolded as to the condition of the country passed through, and not having their eyes unbandaged until in the midst of furnaces, rolling mills, and all the activities of a manufacturing center, they are dazzled by what they see." The trouble was they had lost sight of "the real South — that is to say, of ninety-five per cent thereof."

Unlike the New South, Lewis Blair's "real South" was a backward land, a land of wretched poverty for "the six millions of Negroes who are in the depths of indigence," as well as the ninety percent of the whites who had "nothing beyond the commonest necessaries of life," if that. His beloved South was a retarded region of "dilapidated homesteads," of "fenceless plantations," of illiteracy, of chronic underemployment. To prove his contentions and counter those of "devout believers in a new South," he presented comparative statistics on property values, savings, urban growth, manufacturing, and production. Crude as they were as statistics, his figures served to make his point that the South lagged far behind the American procession, that its people were too poor to accumulate savings and industrialize their economy, and that they lived in a quasi-colonial economy of one-crop agriculture producing raw materials and importing manufactures.

As an active businessman and manufacturer, Blair knew that there were "many causes conspiring to the poverty" of the South, and he listed several, concluding with the degraded status of the Negro. "Each of these causes," he wrote, "would greatly retard the prosperity of the South, or indeed of any country, but all of them combined, destructive as they would necessarily be to prosperity, are not as serious and as fatal as the last cause, namely: the degradation of the Negro." It was the most "far-reaching cause of all," for it served "to intensify all the other drawbacks." The slight regard for sanctity of Negro life and civil rights undermined the sanctity of life and civil rights for all. And similarly the miserable standards of housing, health, diet, education, and morals imposed on the degraded race dragged down the standards of the whole population, whites included, in all these vital areas. The only way out for the South, therefore, was the elevation of the Negro — the immediate elevation of the race.

True to his "hard doctrine" of realism, Blair denied that he rested his argument on the demands of justice, morality, or religion — however strong those demands might be. His was the argument of self-interest based "simply on economic ground, on the ground of advantage to the whites." The Southern white man must be convinced that the only way out of economic stagnation and poverty lay in the elevation of the Negro — the radical and drastic elevation of the whole race. This could not be left to slow evolution. "Man's life is now too short to wait for the natural process of time." The white man "must hasten nature and take a hand himself." The remedies were of a heroic order, for the Negro "must economically, morally, and socially be born again, and self-respect, hope and intelligence are the trinity that will work out his elevation, and they are also the rule of three to work out our own material regeneration."

But if the Negro were to remain "a despised and degraded creature, speaking with bated breath and bowing with head uncovered," if he were "to remain forever a 'nigger,' an object of undisguised contempt, even to the lowest whites," how could he be expected to develop the self-respect, the hope, the ambition that were essential to striving and effort, self-denial and self-discipline? Why should he be expected to become anything but a clown, a drunkard, and a thief? The South was "a veritable land of caste," and so long as the caste system prevailed, the Negro was doomed to servility and humiliation.

Back of the caste system was the myth of white supremacy, and for the pretensions of this dogma Blair had nothing but scorn and contempt. From his reading of history he concluded that "our remote ancestors were as scurvy a lot as ever scourged the earth." Even as late as the Middle Ages, they were "a roving set of pirates and freebooters, whose lives were spent in ravishing and then murdering women, in slaying infants and old men, and in reducing to slavery all able-bodied men who escaped the edge of the sword." Those who chanted the praises of white supremacy might do well to remember that they were "the descendants of these monsters."

The Virginia cynic had not entirely freed himself from the racial credo of his times, for he thought of Negroes as "children" and attributed what progress they had made in America to contact with whites. But on the other hand he rejected the idea of innate or biological inferiority of the race. He was an environmentalist and explained the African's cultural backwardness by reference to "natural surroundings" and "artificial surroundings," by which he meant geographical and cultural environment. Even in Africa, environment had produced a wide variety of cultures, some strong and warlike. He wondered "if the surroundings had been reversed whether we would not now be found at the bottom and the blacks at the top of civilization."

Admittedly race prejudice was deeply entrenched among white Americans, but it was "always a weakness," in extreme form "a badge of dishonor," and it must be eradicated. It was nonsense to hold that it was ineradicable, for it had already been abandoned by advanced nations. In England and France and in Latin nations, generally the Negro was "under no political or social ban," and in Brazil he was "accorded full social and political equality." In fact, a Virginian who had served as United States consul in Rio in the 1850's assured Blair that "he had danced with Negroes at parties and official receptions" and that many of them "were much more elegantly cultured than he was himself." Yet in our "land of the free," a man with the slightest trace of Negro blood, even one who had "rendered great services to his country or to humanity" and who had been "honored in England, France, and Germany," could never feel safe from "snubs, insults, or even kicks from the superior whites." Such conduct was unworthy of a great nation.

The Virginian was perfectly aware that his people, especially the patricians among them, often prided themselves on acts of kindness and deeds of genuine helpfulness toward individual Negroes and favored families and groups. "But we forget," he wrote, "that our kindness to the Negroes proceeds from the standpoint of condescension, and of assumed caste superiority, and we expect it to be received with humility and from a feeling of acknowledged caste inferiority; and if not so received by the Negroes, they are thought impudent and impertinent, and the foundation of our kindness soon dries up."

The Negro was not deceived by the whites. He knew that "this kindness springs mainly from the same benevolence that prompts consideration for their horses and cattle." He knew it was the reward for quiet and complete submission. He remembered very well that slaves had been treated "pretty much as cattle," that the whites had fought a long and desperate war to keep the Negro enslaved, that they had sought to nullify freedom, that they had opposed Negro enfranchisement, "that civil rights were conferred in spite of all their efforts, and that generally they have opposed everything" tending to the advancement of Negroes.

On the surface the Negro appeared acquiescent and contented and the South peacefully and quietly adjusted in its race relations. But the appearance was deceptive and the adjustment could not continue indefinitely. Shortsighted selfishness and an ancient and modern record of injustice had "raised up an enemy, silent and sullen, at our very doors." The policies of white domination, notwithstanding all paternalistic benevolence and charity and condescension, had produced "in the hearts of six millions of fellow-citizens a vast mass of smoldering enmity and bitterness, only awaiting a favorable opportunity to display itself."

If paternalistic benevolence, condescending kindness, the restraints and indulgences of *noblesse oblige,* and all the charities and back-door integration and half-measures were not the answer, what then was the answer? To Lewis Blair the answer was a complete end to segregation and to all forms of discrimination, favoritism, and exclusion on the ground of race or color. He put it plainly:

The Negro must be allowed free access to all hotels and other places of public entertainment; he must be allowed free admittance to all theatres and other places of public amusement; he must be allowed free entrance to all churches, and in all public and official receptions of president, governor, mayor, etc., he must not be excluded by a hostile caste sentiment. In all these things and in all these places he must, unless we wish to clip his hope and crush his self-respect, be treated precisely like the whites, no better, but no worse.

If this seemed impossible to his fellow Southerners, he called upon them to perform an exercise of the imagination they had never tried before: "Let us put ourselves in the Negro's place," he proposed.

Let us feel when passing good hotels there is no admission here; we dare not go in lest we be kicked out; when entering a theatre to be told to go up in the top gallery, no seat for you in the parquette; when entering an imposing church to be told rudely no place in God's house for you, unless there be a gallery, to which you are sent with indifference; or when seeking to attend an official reception, *your* governor's for instance, to be told brusquely no admittance; and suppose this treatment is continued year after year, and as far as we can see is likely to continue *ad infinitum,* would we not have our pride cut to the quick, and would not our morale be greatly lowered, and would we not be greatly handicapped in all our efforts to get along?

But what of the argument that "theatres, hotels and churches are private property, and that to compel them to receive Negroes on equal terms with whites would be to correct one wrong . . . by committing another." His answer was that although they may be private property, "they are public as regards their creation and their functions, and they are of the nature of railways, which may be private property, but which are public institutions." They were licensed by "the public, which means not some, but all the people, not whites alone, but whites and blacks." They could "as properly refuse accommodation to all whose noses indicate a Semitic origin, all whose names or 'rich brogue' betray Hibernian descent . . . as to refuse similar accommodation to all whose faces are black." To shut the doors of any one of these institutions — whether church or theatre or governor's mansion — "in the faces of any portion of the community is to degrade and to humiliate it."

There was one right, held Blair, that was "the right preservative of all rights." That was the ballot, and it was "as absolutely essential for freedom as is the atmosphere for life." Yet it was denied to a great number of Negro citizens of the lower South, and had been since the overthrow of Reconstruction. This was done sporadically by intimidation and fraud rather than consistently by legal disfranchisement, as it was later to be done. The rationalization of Negro disfranchisement was that it was necessary to prevent Negro domination and assure pure government. In Virginia, North Carolina, and Tennessee at the time Blair was writing, however, the Negroes voted freely and fully. Yet these three states showed "greater progress and prosperity" than the other Southern states, "and certainly as much moral and intellectual development." During the administration of the Readjuster party, Virginia "has so-called Negro rule, but the Commonwealth survived, and, in the opinion of many, was much benefited." Protracted one-party rule in Virginia, as in other Southern states, had produced nepotism, corruption, and numerous treasury defalcations. "The fear of Negro rule in the sense of the alarmists," declared Blair, "is a wild and pernicious chimera," and those "who go around wearing the frightful scare-face" in the name of good government were "doing an untold amount of evil to the South." They not only alienated their fellow citizens, but they put a powerful lever in the hands of sectional foes. "The South would make a tremendous ado . . . if a Northern oligarchy of half the population were to claim and assume the right to vote for the whole population." If the South itself continued to do the same thing, it invited a second Reconstruction. "Better surrender now with the honors of the war, or rather with the honors of right, than to wait for years and then surrender at discretion," he warned.

In tackling the racial problem in the public schools, Blair knew that he faced one of the most sensitive areas of white prejudice and fear. He had remedies to propose, he admitted, "that will clash with all of our preconceived ideas, that will be distasteful and repugnant to our prejudices, but," he stoutly maintained, "not to reason and justice." Therefore, to say that reforms were "distasteful and repugnant is really to say nothing against them," for the whole history of progress and the rise of Christianity itself were "simply an overcoming of the violent prejudices and repugnances of the civilized world." Thus fortified by reason and justice, the intrepid Virginia rationalist set forth his revolutionary proposal:

> The remedy proposed is not a bread pill or some soothing syrup, but is a radical and far-reaching one, and is no less than the abandonment of the principle of separate schools, which principle is an efficient and certain mode of dooming to perpetual ignorance both whites and blacks in thinly settled sections.

This was a reform from which even the statesmen of Radical Reconstruction in full possession of power had shied away. And Blair admitted that unless he could convince the whites of the necessity and advantage of the reform, they would "never consent to coeducation [of the races], but will prefer to remain ignorant." He therefore marshaled a wide array of financial, practical, moral, and psychological arguments. Characteristically, he put forward practical reasons of self-interest first, but the real weight of his argument rested upon moral and psychological reasons — some of them quite prophetic and in advance of his time.

The practical and financial arguments were telling. The South was the poorest region of the country with the largest number of children per adult to educate. Its population was not concentrated in towns and cities but widely scattered and dispersed. On top of these enormous burdens was the self-imposed handicap of trying to maintain two separate school systems for the two races, with two corps of teachers and two sets of plants and equipment. The result was a miserably inadequate system, with poor and underpaid teachers, neglected and poorly equipped schools, ill-schooled children, and growing illiteracy. "To fight the battle of education with our present forces and present system of separate schools seems well-nigh hopeless," concluded Blair. To integrate the schools was to relieve the people of part of their burden and improve the schools for both races.

In the second place, separate or segregated schools were morally and psychologically harmful to the children of both races. The damage to the Negro children was most obvious:

> Separate schools are a public proclamation to all of African or mixed blood that they are an inferior caste, fundamentally inferior and totally unfit to mingle on terms of equality with the superior caste. That this is not a temporary and ephemeral but a fundamental and caste inferiority is proven by the fact that opposition does not cease when the temporary inferiority ceases, but still operates, however cultured and refined the Negro may be. Hence it follows that separate schools brand the stigma of degradation upon one-half of the population, irrespective of character and culture, and crush their hope and self-respect, without which they can never become useful and valuable citizens.

The feeling of inferiority "thereby taught the blacks cultivates feelings of abasement and of servile fear of all whom they consider superior — sentiments totally destructive of manliness, courage and self-respect."

The damage segregated schools did to white children was less apparent but no less important:

> Separate schools poison at its very source the stream whence spring the best and noblest fruits of education. . . . At the fountain of education the doctrine of caste, which elsewhere is being successfully combatted, is enshrined in fresh vigor and authority, and it seizes with its rigid, icy grasp the impressible minds of the children, and taints them; and the blind superiority thereby inculcated fosters sentiments of false pride, disregard of the rights of others, and unfeeling haughtiness to all, regardless of color, whom they deem inferiors.

If segregated schooling robbed the Negro children of incentive for achievement, it had the same effect on white children, though from an opposite cause. Since it taught them that "superiority consists in a white skin, they will naturally be satisfied with that kind of superiority, and they will not willingly undergo the tedious, painful and patient ordeal requisite to prepare them for superiority in science, art, literature."

Integrated schools would "emancipate us from this fallacy" and teach both races that "the difference between man and man is not color, but character and conduct." They would dissipate the spirit of "oligarchy, caste, vassalage," and disseminate "correct ideas of personal liberty and equality." They would help remove the brand of degradation from the Negro and the false assumption of superiority from the white and give new incentive for achievement to both.

Those who protested that integration would demoralize the white children "overlook the fact that from earliest childhood they have been subjected to intimate Negro association" with playmates and "the unrestrained influences of Negro nurses at the very time the mind and the heart are most susceptible to influence of every kind." If segregation were the salvation of the whites, it came too late. As for the dangers of demoralization, "History, fiction, the drama, everyday life, all abundantly illustrate the demoralizing effects of the higher upon the lower walks of society. . . . And so it is in the South in the intercourse of the two colors. . . . Demoralization, indeed!"

Looking to the future, Blair saw three alternatives open to the South. The first was to assure Negroes "the whole one hundred per cent" of their rights, so they would be "as free and as equal citizens as the proudest whites." That, he strongly urged, was the way to "peace, happiness, prosperity for all." The second alternative lay in "completely disarming the blacks and reducing them to a condition of complete subordination and degradation." The third lay in half-measures, compromises, and inadequate concessions. Both of the latter alternatives led to "strife, sorrow, adversity for all."

A full retreat to the past, to a cordon sanitaire and an intellectual barricade against criticism and ideas such as the slave regime maintained, was no longer really open to the South. The outside world was now looking over its shoulder and could not be shut out or put off. "We can no more defend our attitude toward the Negroes," he wrote, "than could the Algerian corsairs defend their attitude to the Christian world." The age of caste and privilege was over, for this was "the age of reason." He did not expect to see the walls of caste come tumbling down overnight. "The battle will be long and obstinate," and there would be "many difficulties, delays, and dangers." Nevertheless, he was confident of victory in the end. "We older ones will not see that day," he said, "but our grandchildren will, for the light of coming day already irradiates the eastern sky."

As a final word, Blair addressed an admonition and an appeal to the North. Careful as always to appeal to the motive of self-interest, he pointed out that "in a common country one great section cannot languish without the other sections, even the wealthy and prosperous manufacturing sections, suffering also." The North had a vested interest in the welfare of the South, "and *if* the prosperity of the South is dependent upon the elevation of the Negro, your prosperity is intimately associated with that of the South," and could not "escape the penalty of the South remaining in a stagnant or declining condition in consequence of the Negroes remaining in a state of degradation." In Blair's opinion, "the greatest impediment in the way of Southerners being willing to elevate the Negro" was the North's own "dereliction of duty toward him." This put into the mouth of Southerners the *argumentum ad hominem,* "the argument that you do so yourself." And so every valid criticism of Southern injustice was nullified by an equally valid charge of Northern hypocrisy.

The Virginian was correct in saying the Negro fared better in the North, but knowing less of that section than his own, he conceded too much in saying that "the Negro can travel anywhere without question, and can, with exceptions, attend churches, theatres, and official receptions, and put up at hotels, without fear of affront." He was on firmer ground in his charges of segregated housing, wretched slums, and exclusion of the Negro from skilled and high-wage jobs. "Caste pursues and cripples the Negro in the North as it does in the South." He called on the North to "clear its skirts of the charge of hypocrisy," for until it did so "the seed it sows may be good, but it will fall upon hard and stony soil."

But what was one to say to Southerners who maintained that such radical changes as he proposed were impossible, that they did too much violence to tradition and to history, that they were too great a change and too sudden a break with the past? Lewis Blair had somehow acquired a feeling for the changing fortunes of history, the inevitability of change, and the ephemerality of institutions. He suggested to Southerners that they had less reason than other Americans to talk so confidently about the imperishability of institutions, the inviolable continuity of history, and the improbability of cataclysmic change. Yankees had more excuse to talk in such fanciful language, for they had been spared the tragic experience of history that the South — and most of the world — had suffered, and they could afford to cherish all sorts of illusions about history that were denied Southerners.

And so the Virginian advised Southerners that they

> should be very chary of saying anything is impossible, or that such a thing shall never be, in view of what has taken place and in view

of what we have agreed to in the past quarter-century. In that time we have seen slaves made freedmen at the scratch of one man's pen; by a simple "Be it enacted" we have seen freedmen made citizens, legislators, jurors, teachers of youth, etc., etc.; we have seen civil and political society completely changed, and social life profoundly modified; in a word, we are living in a new world, and we have assented to all these fundamental changes that have made the new world.

Why then this sense of dismay and shock over a new era of innovation and change? Such changes as integrated schools for the 1890's would not be "half so shocking" as were the innovations of the 1860's — still fresh in the memory of many — for they included "Negro emancipation, Negro voting, Negro lawmaking, Negroes sitting on juries, Negroes riding in rail and street cars, our lordly selves standing the meanwhile, Negroes sleeping in the same berths on Pullman cars, etc., etc.; but where is our dread of them?" asked Lewis Harvie Blair in 1889. He asked this question rhetorically, at least in part no doubt, but answered it himself: "Dissipated by experience."

That he could have asked such a question and furnished such an answer (and in good faith, I believe) on the eve of the Jim Crow era, when the Negro was so rapidly to disappear from the polls, the legislatures, the juries, the railroad cars, the streetcars, and the Pullman cars, is not without value to the collector of ironies. It is less a reflection upon his sense of history, however, than upon his gifts for prophecy. We have discovered something worthy of praise in the Virginian's sense of history. And if we can afford to be so generous as to allow him somewhat better than three-quarters of a century as a reasonable margin of error, we might yet conceivably award him honors as a minor prophet.

But that is not the end of the story of Lewis Blair as a prophet of race relations. Nor is it the full measure of its irony. Unfortunately, he lived on through the era of reactionary racism and Jim Crowism. More unfortunate still, there exists unimpeachable evidence that the prophet himself was swept up in the storm of reaction he failed to predict and could not foresee. As late as 1898, he was still reaffirming his equalitarian heresies and lamenting the fact that the South seemed more determined than ever to keep the Negro down and that the North was more and more indifferent to his plight. Blair's great change came sometime after that, between 1898 and his death in 1916. Exactly when and exactly why are unknown. But in his private papers (still in private hands), Mr. Wynes has recently found a manuscript of 270 pages, untitled, unsigned and undated, but unmistakably identified by the handwriting as Blair's own. It is a complete and unqualified recantation of his equalitarian and liberal position of 1889 with regard to the Negro. More than that, it is a total reversal of his earlier stand. It will be recalled that in his book he had indicated three alternatives of racial policy open to the South. The first, his own, was "the whole one hundred per cent" of rights and equality; the second, "complete subordination and degradation"; the third, half-measures, compromises and limited concessions. The two latter he had prophesied led to "strife, sorrow, adversity for all." In the undated handwritten manuscript, belatedly rejecting the first, he now chose not the third but the second alternative; he declared that "the only logical position for the Negro is absolute subordination to the whites." Blair's new logic called for repeal of the Fourteenth and Fifteenth Amendments as well as the complete disfranchisement and total segregation of the Negro. He should be treated kindly but always as an inferior creature — permanently and inherently inferior to whites.

The charitable impulse would be to attribute the change to senility, but the fact is that Blair maintained a lively and intelligent sympathy with Wilsonian progressivism up to the end of his days. His new racial views were quite reconcilable with the progressivism of that day, of course, but neither was the result of senility. Another possibility is that his second marriage in 1898, at the age of sixty-four, to a woman half his age who did not share his racial views may have influenced him. But to those who are familiar with this obscure era of Southern history

Blair's complete reversal will not seem so strange or unpreced-ented. Other examples will come to mind. The most prominent were the Southern Populists, who swung from an advanced brand of racial justice (though more limited than Blair's) to an extreme brand of racial injustice — with Tom Watson of Geor-gia as the classic instance. Blair only proved it could happen in Virginia too, on the other side of the railroad tracks and in one of the finer mansions. The recantation was never published, but there it is, a sad commentary on the frailty of mind and the infirmity of principle.

It is hard to say now which is the greater biographical enigma, the ringing affirmation of faith in 1889 or the silent recantation of old age. But across the dark years of reaction, years of "strife, sorrow, adversity" that he himself predicted would follow the course he later took, the boldly prophetic pronouncement of '89 stands forth as clear, as relevant, and as challenging as it was when it was published more than seventy-five years ago.

8

The National Decision Against Equality

THE state of mind in Louisiana on the accommodation of races on the eve of the Jim Crow era was rather similar to that in Virginia in 1889, when Lewis Harvie Blair published his heretical brief for Negro equality. There were significant differences, to be sure, as will appear, but the similarities were there also. "Perhaps the most striking aspect of race relations in Louisiana from 1877 to 1898," according to a recent study, "was the absence of system. There existed no consistent, thorough, and effective system of social control, legal or extralegal, governing relations between the races. The place of the Negro and his relationship to the white man had yet to be carefully defined." [1] There was absolutely no doubt that blacks were subject to a great deal of discrimination, political manipulation, economic coercion, and brutal treatment. On the other hand, there is no question that there were exceptions to the rule, that oppression had not been systematized or legalized, and that a degree of variety and experiment still prevailed in race relations, partly the heritage of slavery, partly of Latin culture, and partly of Reconstruction. Among whites there was still enough flexibility amid conformity to permit debate on issues that were soon to be completely closed to discussion. George W. Cable, a native Louisianian, could publish views on Negro rights that were not quite

[1] Henry C. Dethloff and Robert R. Jones, "Race Relations in Louisiana, 1877–1898," *Louisiana History*, IX (Fall, 1968), 305.

so heretical as those that the Virginian Blair was publishing about the same time, but they would soon appear wildly heretical.[2] On the black side there remained enough resources of leadership, civil rights, and political organization to support considerable protest and resistance to white aggression.

Resources of black political power in Louisiana were not negligible. The number of registered Negro voters in the state increased from 1868 until 1898, and as late as 1890 they outnumbered whites on the voting rolls, though the two races were roughly equal in population.[3] There was a certain amount of sham in the disparity of black voter registration, for in some of the cotton plantation parishes, large numbers were fraudulently registered and corruptly voted to swell Democratic election returns.[4] Nevertheless, Negro votes were in great demand, and Negroes capitalized on the situation to win offices, favors, and recognition from both parties. Black politicians regularly won seats in the state legislature, eighteen of them in the one sitting in 1890. A few parishes elected black sheriffs, and some seated black school board members and other officials. Negroes frequently served on juries, sometimes bid successfully on state construction contracts, and at times made their weight felt in organized labor. They were generally, though not invariably, excluded or segregated in hotels, restaurants, theaters, schools, and libraries. Social contact between the races persisted, particularly at New Orleans, in a variety of sports, on the beaches of Lake Pontchartrain, and in some churches and bars. Two efforts to pass antimiscegenation failed in the 1880's and another in 1892, and newspapers continued to report interracial marriages. No uniformity of race relations existed in public transportation. Segregated steamboat lines paralleled integrated railroad lines, but while practices differed on both, "integration seems to have

[2] George W. Cable, *The Silent South* (New York, 1885) and *The Negro Question* (New York, 1890).

[3] Dethloff and Jones, "Race Relations in Louisiana," 306, 308.

[4] William Ivy Hair, *Bourbonism and Agrarian Protest: Louisiana Politics, 1877–1900* (Baton Rouge, 1969), 115.

been the rule rather than the exception on most Louisiana railroads." What discrimination and segregation existed were carried out on the responsibility of private owners or local managers and not by requirement or authority of the law. That was to come later. Limited as were the rights and status of the black man in this period, they were to undergo drastic deterioration in the era that followed.[5]

One source of leadership and strength that Louisiana Negroes enjoyed and that blacks of no other state shared was a well-established upper class of mixed racial origin in New Orleans with a strong infusion of French and other Latin intermixtures. Among these people were descendants of the "Free People of Color," some of them men of culture, education, and wealth, often with a heritage of several generations of freedom. Unlike the great majority of Negroes, they were city people with an established professional class and a high degree of literacy. Their views found expression in at least four Negro newspapers that existed simultaneously in New Orleans in the 1880's. By ancestry as well as by residence, they were associated with Latin cultures that were in some ways at variance with Anglo-American ideas of race relations. Their forebears had lived under the Code Noir decreed for Louisiana by Louis XIV, and their city faced out upon Latin America. This group had taken the lead in fighting for Negro rights during Reconstruction and were in a natural position to resist the tide of legal segregation. When it touched their shores they were the first to speak out.

By 1890 only three states, Florida, Mississippi, and Texas, had Jim Crow laws requiring railroads to carry Negroes in separate cars or behind partitions. If Louisiana lagged slightly behind the leaders of reaction in this respect, little of the credit would seem to be due to the influence of the conservative white rulers of the state and the restraining effects of paternalism and

[5] Dethloff and Jones, "Race Relations in Louisiana," 307–17. For a somewhat different reading of the record see Roger A. Fischer, "Racial Segregation in Ante Bellum New Orleans," *American Historical Review*, LXXIV (February 1969), 937.

noblesse oblige often imputed to their class. Their record of broken pledges for the protection of Negro life, liberty, and property was one of the worst in the South. In lynchings, Louisiana ranked third and in public education last among the states. In 1887, the state government had smiled upon a bloody repression of striking sugar workers that took at least thirty lives and, in the years following, tolerated vigilante campaigns against black strikers.[6] When the propaganda for legalizing and requiring racial segregation got under way, the leading newspaper spokesmen of the conservative regime applauded the movement instead of resisting it.[7]

On May 24, 1890, a few days after the railroad segregation bill was reported in the legislature, that body received a memorial entitled "Protest of the American Citizens' Equal Rights Association of Louisiana Against Class Legislation." Signed by a committee of seventeen members of the Association, all apparently Negroes, the memorial denounced the proposed separate-car law as "unconstitutional, un-American, unjust, dangerous and against sound public policy." Such a law, the protest continued, would be "a free license to the evilly disposed that they might with impunity insult, humiliate and otherwise maltreat inoffensive persons, and especially women and children who should happen to have a dark skin." [8]

Two of the signers of the memorial, Louis A. Martinet and Rudolphe L. Desdunes, colored members of the French-speaking community of New Orleans, were to figure prominently in carrying the fight against the law to the United States Supreme Court. The son of a Creole father and a slave mother, Martinet was a young attorney and physician of the city who had been a

[6] Hair, *Bourbonism and Agrarian Protest,* 170–97.

[7] New Orleans *Times-Democrat,* July 9, 1890; New Orleans *Picayune,* July 10, 1890.

[8] This document and several others pertinent to the *Plessy* v. *Ferguson* Case have been ably edited and recently published by Otto H. Olsen (ed.), *The Thin Disguise: Turning Point in Negro History, Plessy v. Ferguson, A Documentary Presentation, 1864–1896* (New York, 1967). For the "Memorial" quoted, see pp. 47–50. Such documents as appear in Olsen's volume will be cited there as the most convenient reference.

Democrat but had recently gone over to the Republicans. In 1889 he founded the New Orleans *Crusader,* a militant weekly paper devoted to the cause of Negro rights. Desdunes, descendant of free Negroes of New Orleans, was a friend of Martinet, a contributor to the *Crusader,* a poet, and later the author of a history of his people, *Nos Hommes et Notre Histoire,* published in 1911.

Under the prodding of the *Crusader,* the eighteen Negro members of the Legislature, with the aid of railroad interests who opposed the bill, succeeded in stalling the progress of the separate-car legislation for a while. But on July 10, 1890, the Assembly passed the bill, the Governor signed it, and it became law. Rightly or wrongly, Martinet and Desdunes in separate articles published in the *Crusader,* placed a heavy share of the blame for passage of the bill on the colored Republican members of the legislature. According to Desdunes, they had been promised votes against the Jim Crow bill in exchange for votes to overturn the Governor's veto of the unpopular Louisiana Lottery bill and these promises were then broken. "The Lottery bill could not have passed without their votes," wrote Martinet; "they were completely the masters of the situation" had they only withheld their support. "But in an evil moment our Representatives turned their ears to listen to the golden siren," and "in emulation of their white colleagues, they did so for a 'consideration.' "[9]

Putting aside recrimination, Martinet declared: "The Bill is now a law. The next thing is what we are going to do." He thought there was merit in Desdunes' idea of boycotting the railroads, but he was more interested in fighting the case in the courts. "The next thing is . . . to begin to gather funds to test the constitutionality of this law. We'll make a case," wrote the young lawyer, "a test case, and bring it before the Federal Courts on the ground of the invasion of the right of a person to

[9] New Orleans *Crusader,* July 19, 1890; R. L. Desdunes to A. W. Tourgée, February 28, 1892, Tourgée Papers, Chatauqua County Historical Museum, Westfield, New York.

travel through the States unmolested." [10] Nothing came of the proposal for more than a year. Then, on September 1, 1891, a group of eighteen men of color, all but three of them with French names, such as Esteves, Christophe, Bonseigneur, and Labat, including Desdunes and Martinet, formed a Citizens' Committee to Test the Constitutionality of the Separate Car Law. Money came in slowly at first, but by October 11, Martinet could write that the committee had already collected $1,500 and that more could be expected "after we have the case well started." Even before the money was collected, Martinet had opened a correspondence about the case with Albion Winegar Tourgée, of Mayville, New York, and on October 10 the Citizens' Committee formally elected Tourgée "leading counsel in the case, from beginning to end, with power to choose associates." [11]

This action called back into the stream of history a name prominent in the annals of Reconstruction. Albion W. Tourgée was in 1890 probably the most famous surviving carpetbagger. His fame was due not so much to his achievements as a carpetbagger in North Carolina, significant though they were, as to the six novels about his Reconstruction experience that he had published since 1879. Born in Ohio, of French Huguenot descent, he had served as an officer in the Union Army, and moved to Greensboro, North Carolina, in 1865 to practice law. He soon became a leader of the Republican party, took a prominent part in writing the radical Constitution of North Carolina, and served as a judge of the superior court for six years with considerable distinction. On the side he helped prepare a codification of the state law and a digest of cases.[12]

Tourgée's Southern enemies questioned his public morals and his political judgment, but not his intelligence and certainly not

[10] New Orleans *Crusader,* July 19, 1890; also R. L. Desdunes, *Nos Hommes et Notre Histoire* (Montreal, 1911), 183–94.

[11] L. A. Martinet to A. W. Tourgée, October 11, 1891, Tourgée Papers.

[12] Otto H. Olsen, *Carpetbagger's Crusade: The Life of Albion Winegar Tourgée* (Baltimore, 1965), is an excellent study, the best biography of a carpetbagger we have.

his courage. They knew him too well. "Tourgée was a special case," as Edmund Wilson has shrewdly remarked. "He was a Northerner who resembled Southerners: in his insolence, his independence, his readiness to accept a challenge, his recklessness and ineptitude in practical matters, his romantic and chivalrous view of the world in which he was living. . . . And he evidently elicited *their* admiration or he could never have survived as so provocative an antagonist fourteen years, as he did, in their midst." [13] Although he entitled his most successful novel on Reconstruction *A Fool's Errand,* he had by no means lost the convictions that inspired his crusade for the freedmen of North Carolina, and he brought to the fight against segregation in Louisiana a combination of zeal and ability that the Citizens' Committee of New Orleans would have found impossible at that time to equal.

In a period of mounting racism, when former friends of the Negro had died off, grown silent, or changed their views, Tourgée stood out, as his biographer says, as "the most vocal, militant, persistent, and widely heard advocate of Negro equality in the United States, black or white. He was the Garrison of a new struggle," all the more conspicuous because "the times were wrong." In speeches, articles, books, and since 1888 in his weekly column in the Chicago *Inter Ocean,* he kept up a passionate, polemical, and unrelenting attack on the enemies and wrongs of the black men. In October 1891, when the New Orleans committee opened correspondence with him, Tourgée was just beginning a drive to organize a biracial National Citizens Rights Association with himself as provisional president for the defense of Negro rights. Two or three hundred letters a day were arriving from recruits, many of them in the South, and within six months he claimed over one hundred thousand members. George W. Cable of Louisiana, with whom Martinet and his friends were in touch, served on the executive board of Tourgée's association, and a local branch of the NCRA was

[13] Edmund Wilson, *Patriotic Gore: Studies in the Literature of the American Civil War* (Boston, 1962), 537.

soon established in New Orleans. Martinet, speaking for the Citizens' Committee on the Jim Crow law, had ample reason to write Tourgée, "We know we have a friend in you." He informed his friend that the committee's decision electing him their counsel was made "spontaneously, warmly, and gratefully." [14]

To assist Tourgée with local procedure, the committee employed a white Republican lawyer, James C. Walker. Tourgée served throughout without fee. In keeping with a strategy he had in mind, his first suggestion was that the person chosen for defendant in the test case be "nearly white," but that proposal raised some doubts. "It would be quite difficult," explained Martinet, "to have a lady *too* nearly white refused admission to a 'white' car." He pointed out that "people of tolerably fair complexion, even if unmistakably colored, enjoy here a large degree of immunity from that accursed prejudice. . . . To make this case would require some tact." He would volunteer himself, "but I am one of those whom a fair complexion favors. I go everywhere, in all public places, though well-known all over the city, & never is anything said to me. On the cars it would be the same thing. In fact, color prejudice, in this respect, does not affect me. But, as I have said, we can try it, with another." An additional point of delicacy was a jealousy among the darker members of the colored community, who "charged that the people who support our movement were nearly white, or wanted to pass for white." Martinet discounted the importance of this feeling, but evidently took it into account. The critics, he said, had contributed little to the movement.[15]

Railroad officials proved surprisingly cooperative. The first one approached, however, confessed that his road "did not en-

[14] Olsen, *Carpetbagger's Crusade*, 298–317; L. A. Martinet to A. W. Tourgée, October 5, 1891, in Olsen (ed.), *The Thin Disguise*, 55–61; for samples of the 1891–1893 correspondence see Olsen, "Albion W. Tourgée and Negro Militants of the 1890's," *Science and Society*, XXVIII (Spring, 1964), 183–208.

[15] Martinet to Tourgée, October 5, 11, 25, December 7, 1891, Tourgée Papers.

force the law." It provided the Jim Crow car and posted the sign required by law, but told its conductors to molest no one who ignored instructions. Officers of two other roads "said the law was a bad and mean one; they would like to get rid of it," and asked for time to consult counsel. "They want to help us," said Martinet, "but dread public opinion." The extra expense of separate cars was one reason for railroad opposition to the Jim Crow law.[16] It was finally agreed that a white passenger should object to the presence of a Negro in a "white" coach, that the conductor should direct the colored passenger to go to the Jim Crow car, and that he should refuse to go. "The conductor will be instructed not to use force or molest," reported Martinet, "& *our* white passenger will swear out the affidavit. This will give us our *habeas corpus* case, I hope." [17]

On the appointed day, February 24, 1892, Daniel F. Desdunes, son of Louis Desdunes, bought a ticket for Mobile, boarded the Louisville & Nashville Railroad, and took a seat in the white coach. All went according to plan. Desdunes was committed for trial to the Criminal District Court in New Orleans and released on bail. On March 21, Walker, the local attorney associated with Tourgée in the case, filed a plea protesting that his client was not guilty and attacking the constitutionality of the Jim Crow law. He wrote Tourgée that he intended to go to trial as early as he could.[18]

Between the lawyers there was not entire agreement on procedure. Walker favored the plea that the law was void because it attempted to regulate interstate commerce, over which the Supreme Court held that Congress had exclusive jurisdiction. Tourgée was doubtful. "What we want," he wrote Walker, "is not a verdict of not guilty, nor a defect in this law but a decision whether such a law can be legally enacted and enforced in any state and we should get everything off the track and out of the

[16] Martinet to Tourgée, December 7, 1891, Tourgée Papers.

[17] Martinet to Tourgée, December 28, 1891, Tourgée Papers.

[18] Walker to Tourgée, January 2, 9, 21, 1892, Tourgée Papers; New Orleans *Times-Democrat,* March 22, 1892.

way for such a decision." Walker confessed that "It's hard for me to give up my pet hobby that the law is void as a regulation of interstate commerce," and Tourgée admitted that he "may have spoken too lightly of the interstate commerce matter." [19]

However, the discussion was ended abruptly and the whole approach altered before Desdunes' case came to trial by a decision of the State Supreme Court handed down on May 25. In this case, which was of entirely independent origin, the court reversed the ruling of a lower court and upheld the Pullman Company's plea that the Jim Crow law was unconstitutional insofar as it applied to interstate passengers.[20]

Desdunes was an interstate passenger holding a ticket to Alabama, but the decision was a rather empty victory. The law still applied to intrastate passengers, and since all states adjacent to Louisiana had by this time adopted similar or identical Jim Crow laws, the exemption of interstate passengers was of no great importance to the Negroes of Louisiana and it left the principle against which they contended unchallenged. On June 1, Martinet wired Tourgée on behalf of the committee saying, "Walker wants new case wholly within state limits," and asked his opinion. Tourgée wired his agreement.[21]

One week later, on June 7, Homer Adolph Plessy bought a ticket in New Orleans, boarded the East Louisiana Railroad bound for Covington, Louisiana, and took a seat in the white coach. Since Plessy later described himself as "seven-eighths Caucasian and one-eighth African blood," and swore that "the admixture of colored blood is not discernible," it may be assumed that the railroad had been informed of the plan and agreed to cooperate. When Plessy refused to comply with the conductor's request that he move to the Jim Crow car, he was arrested by Detective Christopher C. Cain, and charged with

[19] Walker to Tourgée, February 25, 26, March 6, 8, 18, 1892, Tourgée to Walker, April 1, 1892, Tourgée Papers.

[20] New Orleans *Times-Democrat,* May 26, 1892.

[21] Telegrams: Martinet to Tourgée, May 26, June 1, 1892; Tourgée to Martinet (no date), Tourgée Papers.

violating the Jim Crow car law. Tourgée and Walker then entered a plea before Judge John H. Ferguson of the Criminal District Court for the Parish of New Orleans, arguing that the law Plessy was charged with violating was null and void because it was in conflict with the Constitution of the United States. Ferguson ruled against them. Plessy then applied to the State Supreme Court for a writ of prohibition and certiorari and was given a hearing in November 1892. Thus was born the case of *Plessy* v. *Ferguson.*[22]

The court recognized that neither the interstate commerce clause nor the question of equality of accommodations was involved and held that the sole question was whether a law requiring "separate but equal accommodations" violated the Fourteenth Amendment. Citing numerous decisions of lower federal courts to the effect that accommodations did not have to be identical to be equal, the court, as expected, upheld the law. "We have been at pains to expound this statute," added the court, "because the dissatisfaction felt with it by a portion of the people seems to us so unreasonable that we can account for it only on the ground of some misconception." [23]

Chief Justice Francis Tillou Nicholls, who presided over the court that handed down this decision in 1892, had signed the Jim Crow act as Governor when it was passed in 1890. Previously he had served as the "Redeemer" Governor who took over Louisiana from the carpetbaggers in 1877 and inaugurated a brief regime of conservative paternalism. In those days, Nicholls had denounced race bigotry, appointed Negroes to office, and attracted many of them to his party, Martinet among them. Martinet wrote Tourgée that Nicholls in those years had been "fair & just to colored men" and had, in fact, "secured a degree of protection to the colored people not enjoyed before under Re-

[22] New Orleans *Times-Democrat,* June 9, November 19, 1892.

[23] *Ex parte Homer A. Plessy,* 45 La. Ann. 80, Decision by Justice Charles E. Fenner, December 19, 1892, in Olsen (ed.), *The Thin Disguise,* 71–77.

publican Governors." [24] But in November 1892, the wave of Populist radicalism was reaching its crest in the South. Not only were black and white farmers aroused, but New Orleans workers of both races had just shaken the city by a militant general strike. Forty-two union locals with over 20,000 members, who with their families made up nearly half the population of the city, struck for a ten-hour day, overtime pay, union recognition, and a closed shop. Business came to a halt and bank clearings were cut in half. It has been described by one historian as "the first general strike in American history to enlist both skilled and unskilled labor, black and white, and to paralyze the life of a great city." The governor of the state proclaimed martial law, and under threat of force the strikers agreed to a weak compromise.[25] This was the immediate background of the court's decision, though it would probably have gone the same way anyhow. But the course of Judge Nicholls since 1877 typified the concessions to racism that conservatives of his class were making — and had been making — to divert white farmers and workers from their course of rebellion.

Tourgée and Walker were denied a rehearing but obtained a writ of error, which was accepted by the United States Supreme Court.[26] There ensued a delay of three years before the Supreme Court got around to hearing the case argued. The delay, oddly enough, pleased Tourgée. What he most feared was an unfavorable (and irreversible) decision. He was convinced that time worked in his favor and that with more years of agitation and crusade by the new National Citizens Rights Association, the

[24] Barnes F. Lathrop (ed.), "An Autobiography of Francis T. Nicholls, 1834–1881," *Louisiana Historical Quarterly,* XVII (1934), 246–67; New Orleans *Daily Picayune,* June 27, 1877; Martinet to Tourgée, October 5, 1891, Tourgée Papers.

[25] Roger W. Shugg, "The New Orleans General Strike of 1892," *Louisiana Historical Quarterly,* XXI (1938), 547, 539.

[26] Assignment of Errors, *ex parte Homer A. Plessy,* January 5, 1893 (Supreme Court Records, National Archives) in Olsen (ed.), *The Thin Disguise,* 74–77; Tourgée to Martinet, October 31, 1893, *ibid.,* 78–80.

tide of public opinion could be turned. The tide, of course, continued to mount against his cause. The brief that Albion Tourgée filed with the Supreme Court in behalf of Plessy in October 1895 breathed a spirit of equalitarianism that was more in tune with his carpetbagger days than with the prevailing spirit of the mid-nineties. And it was no more in accord with the dominant mood of the Court than was the lone dissenting opinion later filed by Justice John Marshall Harlan, which echoed many of Tourgée's ringing phrases.

At the very outset, however, Tourgée advanced an argument in behalf of his client that unconsciously illustrated the paradox that had from the start haunted the American attempt to reconcile strong color prejudice with equalitarian commitments. Plessy, he contended, had been deprived of property without due process of law. The "property" in question was the "reputation of being white." It was "the most valuable sort of property, being the master-key that unlocks the golden door of opportunity." Intense race prejudice excluded any man suspected of having Negro blood "from the friendship and companionship of the white man," and therefore from the avenues to wealth, prestige, and opportunity. "Probably most white persons if given the choice," he held, "would prefer death to life in the United States as colored persons."

Since Tourgée had proposed that a person who was "nearly white" be selected for the test case, it may be presumed that he did so with this argument in mind. He doubtless hoped thereby to appeal to the preferential treatment the Supreme Court notoriously gave to property rights. Of course, this was not a defense of the colored man against discrimination by whites, but a defense of the "nearly" white man against the penalties of color. From such penalties the colored man himself admittedly had no defenses. The argument, whatever its merits, apparently did not impress the Court.

Tourgée went on to develop more relevant points. He emphasized especially the incompatibility of the segregation law with the spirit and intent of the Thirteenth and Fourteenth Amendments, particularly the latter. Segregation perpetuated distinctions "of a servile character, coincident with the institution of slavery." He held that "slavery was a caste, a legal condition of subjection to the dominant class, a bondage quite separable from the incident of ownership." He scorned the pretense of impartiality and equal protection advanced in defense of the "separate but equal" doctrine. "The object of such a law," he declared, "is simply to debase and distinguish against the inferior race. Its purpose has been properly interpreted by the general designation of 'Jim Crow Car' law. Its object is to separate the Negroes from the whites in public conveyances for the gratification and recognition of the sentiment of white superiority and white supremacy of right and power." He asked the members of the Court to imagine the tables turned and themselves ordered into a Jim Crow car. "What humiliation, what rage would then fill the judicial mind!" he exclaimed.

The clue to the true intent of the Louisiana statute was that it did not apply "to nurses attending the children of the other race." On this clause he observed:

> The exemption of nurses shows that the real evil lies not in the color of the skin but in the relation the colored person sustains to the white. If he is a dependent, it may be endured: if he is not, his presence is insufferable. Instead of being intended to promote the *general* comfort and moral well-being, this act is plainly and evidently intended to promote the happiness of one class by asserting its supremacy and the inferiority of another class. Justice is pictured blind and her daughter, the Law, ought at least to be color-blind.[27]

Looking to the future, Tourgée asked, "What is to prevent the application of the same principle to other relations" should the separate-car law be upheld? Was there any limit to such laws?

> Why not require all colored people to walk on one side of the street and the whites on the other? Why not require every white man's

[27] This passage is from Tourgée's brief filed with the Court April 6, 1896, a copy of which is in the Tourgée Papers. Other quotations are from his brief filed in the October term, 1895, most of which is reprinted in Olsen (ed.), *The Thin Disguise,* 80–103.

house to be painted white and every colored man's black? Why may it not require every white man's vehicle to be of one color and compel the colored citizen to use one of different color on the highway? Why not require every white business man to use a white sign and every colored man who solicits customers a black one? One side of the street may be just as good as the other and the dark horses, coaches, clothes and signs may be as good or better than the white ones. The question is not as to the *equality* of the privileges enjoyed, but the *right of the State to label one citizen as white and another as colored* in the common enjoyment of a public highway as this court has often decided a railway to be.

Two other briefs in support of Plessy's case, both by Southern whites out of the Radical Reconstruction past, were laid before the court. One was by James C. Walker of Louisiana, Tourgée's associate in the case. The other was written by Samuel F. Phillips of North Carolina, an old friend and onetime Scallawag colleague of onetime carpetbagger Tourgée. Thirteen years before the *Plessy* decision Phillips, then United States Solicitor General, had suffered defeat before the same court in the *Civil Rights Cases* of 1883.[28]

The Supreme Court did not hand down a decision on *Plessy* v. *Ferguson* until 1896. In the four years that had intervened since Homer Plessy was arrested in New Orleans, the South had quickened the pace of retreat from its always reluctant commitment to equality and the Fourteenth Amendment, and it had met with additional acquiescence, encouragements, and approval in the North. New segregation laws had been adopted. Lynching had reached new peaks. Frightened by Populist gains in 1892 and 1894, Southern conservatives raised the cry of Negro Domination and called for White Solidarity. Two states had already disfranchised the Negro, and several others, including Louisiana, were planning to take the same course. In New Orleans, Louis Martinet's valiant *Crusader* had folded and his forces

[28] Brief for Homer A. Plessy by S. F. Phillips and F. D. McKenney, *File Copies of Briefs, 1895*, VIII (October Term, 1895), in Olsen, *The Thin Disguise*, 103–108.

were in disarray. In 1892, Congress defeated the Lodge Bill to extend federal protection to elections, and in 1894, it wiped from the federal statutes a mass of Reconstruction laws for the protection of equal rights. And then, on September 18, 1895, Booker T. Washington delivered a famous speech embodying the so-called "Atlanta Compromise," which was widely interpreted as an acceptance of subordinate status for the Negro by the foremost leader of the race.

Given the strong tide of reaction in public opinion on race relations, the weakness of white friends of the Negro, and the seeming acquiescence of the Negro himself, it may well have appeared to the court that it was not within the capabilities of the judicial process to stem the tide — even if the court had been so disposed. If in this case, however, the court acted out of fear of public opinion or deference to majority will, it had certainly shown no such timidity in defying public opinion the previous year, 1895, when it handed down a succession of extremely unpopular opinions in defense of property rights. By 1896 the court may not have been able to stem the tide of segregation, as it might have earlier. On the other hand, it was under no obligation or necessity to give impetus to reaction by the rhetoric in which it couched its opinion.

On May 18, 1896, Justice Henry Billings Brown, of Michigan residence and Massachusetts birth, delivered the opinion of the court on the case of *Plessy* v. *Ferguson*. His views upholding the separate-but-equal doctrine were in accord with those of all his brothers, with the possible exception of Justice Brewer, who did not participate, and the certain exception of Justice Harlan, who vigorously dissented. In approving the principle of segregation, Justice Brown was also in accord with the prevailing climate of opinion and the trend of the times. More important for purposes of the decision, his views were in accord with a host of state judicial precedents, which he cited at length, as well as with unchallenged practice in many parts of the country, North and South. Furthermore, there were no federal judicial precedents to the contrary.

Whether Brown was well advised in citing as his principal authority the case of *Roberts* v. *City of Boston* is another matter. The fame of Chief Justice Lemuel Shaw of the Massachusetts Supreme Court was undoubtedly great, and in this case he unquestionably sustained the power of Boston to maintain separate schools for Negroes and rejected Charles Sumner's plea for equality before the law. But that was in 1849, twenty years before the Fourteenth Amendment, which, as Tourgée pointed out, should have made a difference. More telling was Brown's mention of the action of Congress in establishing segregated schools for the District of Columbia, an action endorsed by Radical Republicans who had supported the Fourteenth Amendment and sustained by regular Congressional appropriations ever since. Similar laws, wrote Brown, had been adopted by "the legislatures of many states, and have been generally, if not uniformly, sustained by the courts."

The validity of such segregation laws, the Justice maintained, depended on their "reasonableness." And in determining reasonableness, the legislature "is at liberty to act with reference to the established usages, customs, and traditions of the people, and with a view to the promotion of their comfort, and the preservation of the public peace and good order."

In addition to judicial precedent and accepted practice, Justice Brown ventured into the more uncertain fields of history, sociology, and psychology for support of his opinion. The framers of the Fourteenth Amendment, he maintained, "could not have intended to abolish distinctions based upon color, or to enforce social, as distinguished from political, equality." The issue of "social equality" was hardly in question here, but there were certainly grounds for maintaining that the framers of the amendment were under the impression that they intended to abolish all legal distinctions based on color.

The sociological assumptions governing Justice Brown's opinion were those made currently fashionable by Herbert Spencer and William Graham Sumner, but the dictum of Chief Justice Shaw in 1849, that prejudice "is not created by law, and probably cannot be changed by law," can hardly be attributed to the influence of either of those theorists. "We consider the underlying fallacy of the plaintiff's argument," said Brown,

> to consist in the assumption that the enforced separation of the two races stamps the colored race with the badge of inferiority. If this is so, it is not by reason of anything found in the act, but solely because the colored race chooses to put that construction upon it. . . . The argument also assumes that social prejudices may be overcome by legislation, and that equal rights cannot be secured by the negro except by an enforced commingling of the two races. We cannot accept this proposition. . . . Legislation is powerless to eradicate racial instincts, or to abolish distinctions based upon physical differences, and the attempt to do so can only result in accentuating the difficulties of the present situation. If the civil and political rights of both races be equal, one cannot be inferior to the other civilly or politically. If one race be inferior to the other socially, the constitution of the United States cannot put them upon the same plane.[29]

The most fascinating paradox in American jurisprudence is that the opinions of two sons of Massachusetts, Shaw and Brown, should have bridged the gap between the radical equalitarian commitment of 1868 and the reactionary repudiation of that commitment in 1896; and that a Southerner should have bridged the greater gap between the repudiation of 1896 and the radical rededication of the equalitarian idealism of Reconstruction days in 1954. For the dissenting opinion of Justice Harlan, embodying many of the arguments of Plessy's ex-carpetbagger counsel, foreshadowed the court's eventual repudiation of the *Plessy* v. *Ferguson* decision and the doctrine of "separate but equal" more than half a century later.

John Marshall Harlan is correctly described by Robert Cushman as "a Southern gentleman and a slaveholder, and at heart a conservative." [30] His famous dissent in the *Civil Rights Cases* of 1883 had denounced the "subtle and ingenious verbal criticism" by which "the substance and spirit of the recent amendments of

[29] *Plessy* v. *Ferguson,* 163 U.S. 537 (1896).
[30] Robert Cushman in *Dictionary of American Biography,* VIII, 269.

the Constitution have been sacrificed." In 1896, the "Great Dissenter" was ready to strike another blow for his adopted cause.

Harlan held the Louisiana segregation law in clear conflict with both the Thirteenth and Fourteenth Amendments. The former "not only struck down the institution of slavery," but also "any burdens or disabilities that constitute badges of slavery or servitude." Segregation was just such a burden or badge. Moreover, the Fourteenth Amendment "added greatly to the dignity and glory of American citizenship, and to the security of personal liberty," and segregation denied to Negroes the equal protection of both dignity and liberty. "The arbitrary separation of citizens, on the basis of race, while they are on a public highway," he said, "is a badge of servitude wholly inconsistent with the civil freedom and the equality before the law established by the constitution. It cannot be justified upon any legal grounds."

Harlan was as scornful as Tourgée had been of the claim that the separate-car law did not discriminate against the Negro. "Every one knows," he declared, that its purpose was "to exclude colored people from coaches occupied by or assigned to white persons." This was simply a poorly disguised means of asserting the supremacy of one class of citizens over another. The justice continued:

> But in view of the constitution, in the eye of the law, there is in this country no superior, dominant, ruling class of citizens. There is no caste here. *Our constitution is color-blind,* and neither knows nor tolerates classes among citizens. In respect of civil rights, all citizens are equal before the law. The humblest is the peer of the most powerful. The law regards man as man, and takes no account of his surroundings, or of his color when his civil rights as guaranteed by the supreme law of the land are involved. . . . We boast of the freedom enjoyed by our people above all other peoples. But it is difficult to reconcile that boast with a state of law which, practically, puts the brand of servitude and degradation upon a large class of our fellow citizens — our equals before the law. The thin disguise of "equal" accommodations for passengers in railroad coaches will not mislead any one, nor atone for the wrong this day done.

> The present decision, it may well be apprehended [predicted Harlan], will not only stimulate aggressions, more or less brutal and irritating, upon the admitted rights of colored citizens, but will encourage the belief that it is possible, by means of state enactments, to defeat the beneficent purposes which the people of the United States had in view when they adopted the recent amendments of the constitution.

If the state may so regulate the railroads, "why may it not so regulate the use of the streets of its cities and towns as to compel white citizens to keep on one side of a street, and black citizens to keep on the other," or, for that matter, apply the same regulations to streetcars and other vehicles, or to courtroom, the jury box, the legislative hall, or to any other place of public assembly? "In my opinion," concluded the Kentuckian, "the judgment this day rendered will, in time, prove to be quite as pernicious as the decision made by this tribunal in the Dred Scott Case." [31]

The country received the news of the *Plessy* v. *Ferguson* decision with a response that differed from the way it reacted to the decision in the *Civil Rights Cases* thirteen years earlier. In 1883, the news of the court's action had precipitated hundreds of editorials, some indignant rallies, Congressional bills, a Senate report, and much general debate. Otto H. Olsen has made a systematic survey of the national reaction to the *Plessy* decision which indicates that the news was not received without dissent, that hostility was not confined to the Negro press, and that the presumption of consensus about segregation in 1896 is an exaggeration. Nevertheless, it is clear that a great change had taken place since 1883 and that the court gave voice to the dominant mood of the country.[32] Justice Harlan spoke for the convictions of a bygone era.

The racial aggressions that the justice foresaw came in a flood after the decision of 1896. Even Harlan indicated by his opinion of 1899 in *Cummings* v. *Board of Education* that he saw noth-

[31] *Plessy* v. *Ferguson,* 163 U.S. 559–61.
[32] Olsen, *The Thin Disguise,* 25–27.

ing unconstitutional in segregated public schools.[33] Virginia was the last state in the South to adopt the separate-car law, and she resisted it until 1900. Up to that year, this was the only law of the type adopted by a majority of the Southern states. But on January 12, 1900, the editor of the Richmond *Times* was in full accord with the new spirit when he asserted: "It is necessary that this principle be applied in every relation of Southern life. God Almighty drew the color line and it cannot be obliterated. The negro must stay on his side of the line and the white man must stay on his side, and the sooner both races recognize this fact and accept it, the better it will be for both."

With incredible thoroughness the color line *was* drawn and the Jim Crow principle applied — even to areas that Tourgée and Harlan had suggested a few years before as absurd extremes. In sustaining the constitutionality of the new Jim Crow laws, courts universally and confidently cited *Plessy* v. *Ferguson* as the leading authority. They continued to do so for more than half a century.

On April 4, 1950, Justice Robert H. Jackson wrote old friends in Jamestown, New York, of his surprise in running across the name of Albion W. Tourgée, once a resident of the nearby village of Mayville, in connection with segregation decisions then pending before the Supreme Court. "The Plessy case arose in Louisiana," he wrote,

> and how Tourgée got into it I have not learned. In any event, I have gone to his old brief, filed here, and there is no argument made today that he would not make to the Court. He says, "Justice is pictured blind and her daughter, The Law, ought at least to be color-blind." Whether this was original with him, it has been gotten off a number of times since as original wit. Tourgée's brief was filed April 6, 1896, and now, just fifty-four years after, the question is again being argued whether his position will be adopted and what was a defeat for him in '96 be a post-mortem victory.[34]

[33] *Cummings* v. *Board of Education,* 175 U.S. 528 (1899). Consistency was not Justice Harlan's strongest point.

[34] Justice Robert H. Jackson to Ernest Cawcroft and Walter H. Edson, April 4, 1950. Copy in Robert H. Jackson Papers. Professor Paul Freund

Plessy v. *Ferguson* remained the law of the land for exactly fifty-eight years, from May 18, 1896, to May 17, 1954. Then, at long last, came a vindication, "a post-mortem victory" — not only for the ex-carpetbagger Tourgée, but for the ex-slaveholder Harlan as well.

---

of Harvard University called my attention to the letter and kindly sent me a copy.

9

The Strange Career of a Historical Controversy

ONE traditional way we have of arriving at historical truth — or more accurately groping for it — might be described as adversary procedure. The term, like the method, suggests the court of law, with its assumption of opposing if not antagonistic claims and its presumptions that any litigant deserves a hearing for the best case that can be made for his cause. This tradition thrusts historians rather arbitrarily and awkwardly into the role of advocates. Their adjustment has not always been a happy one. In response to Shakespeare's admonition, "Do as adversaries do in law,/Strive mightily, but eat and drink as friends," historians have unfortunately been more likely to follow the first rather than the second part. Too often they have tended to think of the adversary in the manner of Milton when he capitalized the word and left no doubt to Whom he was referring. Rejecting the more genial convention of barristers, historians have too readily fallen in with the lawyer's less fortunate habit of being more concerned about winning the suit than about pursuing the truth.

Examples of adversary procedure abound in our historical literature. So numerous are they and so cooperative the historians in providing examples, that teachers have seized upon them as the meat of history and a method of pedagogy. Publishers have obligingly packaged them in reprint combinations, organizing whole series of books along the adversary-procedure line. So taught and obliged so to read, students have come to perceive history in this manner, and to think and speak in either-or terms of such complicated aspects of American history as priority in the origins of slavery or race prejudice, democratic-or-aristocratic colonial government, motives of the Constitution framers, and the causes of the War of 1812. Students and professors fall readily into making simplistic dichotomies in treating the origins and character of abolitionists, Jacksonian Democrats, Radical Republicans, Scalawags and Carpetbaggers, Mugwumps, Populists, Progressives, New Dealers, and Cold Warriors. The students are not to blame, and it is difficult to say whether the teachers, the publishers, or the historians bear the greatest share of responsibility.

Historians are sometimes trapped unawares in the role of advocate in an adversary procedure. Typically, one of them will hit upon a new way of looking at an old question, or a new answer, a new interpretation, or a new question to ask of old data. Since it *is* new, he will emphasize it, perhaps overemphasize it in his publication simply to overcome the weight of presumption against any challenge of accepted views. In time, allies will rally to his support with reviews, articles, and books that sustain his new findings, facts, interpretations, and strengthen his conviction that he is right. The challenge is on the way to becoming the Accepted View. Should the new view sustain hope for public policies and support for views in public controversy which the author shares, his sympathies may become involved with his vanity as investments in his thesis. Faced eventually with adverse criticism and evidence that contradicts his thesis, he is tempted to assume a defensive posture. He may well become more resentful on seeing his meaning misconstrued or distorted for polemical advantage. And even more so on seeing the young, thoroughly indoctrinated with the notion that adversarial notoriety is the approved ladder to advancement in the guild, join the pack in hot pursuit. In these circumstances, historians of stature have been known to stoop to vituperative exchanges with their opponents in public print. Along with the pursuit of truth, the Shakespearean ideal of genial adversaries is likely to be abandoned, and in its place there comes to prevail something more resembling the concept of the Apostle Peter: "Your adversary, the devil, as a roaring lion, walketh about, seeking whom he may devour."

The instance of adversarial historiography under scrutiny here — the literature on the origins and development of racial segregation in the American South — is a minor and not altogether typical one. The literature of the subject illustrates some, but fortunately not all — and particularly not the recriminatory and vituperative — aspects of the model described above. In fact, the present writer was able to say in summing up the criticisms of his thesis in 1966 that, "Without exception, they have been offered in a spirit of detachment, generosity, and courtesy that I shall try to emulate in making use of their findings and dealing with their criticisms in revising my own work." [1] There is no guarantee, of course, of such continued good fortune in the maintenance of decorum and the attraction of Shakespearean adversaries. Since no end to the controversy yet seems in sight, it might be well to survey its course and if possible lay it to rest before any serious breach in good manners occurs. One purpose in so doing is to see if there is any mutually agreeable reformulation of the problem that will get at its essence better than previous formulations. If so, it might release the historians from their adversarial involvement and obligations and permit them all to retire from the contest with honor. It might also serve to redirect energies for the pursuit of truth that might otherwise be spent in the winning of a suit.

It began in a mild way with a series of lectures in the fall of 1954 that were published the following year. The thesis was relatively simple, though it was guardedly advanced and elabo-

[1] C. Vann Woodward, *The Strange Career of Jim Crow,* second revised ed. (New York, 1966), vii. Subsequent references will be to this edition, but there have been two others, the original in 1955 and a first revision in 1957.

rately qualified, and the implications were *not* simple. Briefly stated, the thesis was, first, that racial segregation in the South in the rigid and universal form it had taken did not appear with the end of slavery, but toward the latter years of the century and later; and second, that before it appeared in this form there transpired an era of experiment and variety in race relations of the South in which segregation was not the invariable rule.

The initial response had little to do with scholarly scruple and research. In this respect the origins-of-segregation question bears a resemblance to another and older example of adversarial historiography, the origins-of-slavery problem. In his account of the latter controversy, Winthrop D. Jordan points out that passions were stirred because if race prejudice were a consequence of slavery, the hope for eliminating prejudice was better than if slavery were a consequence of prejudice.[2] Similarly the thesis about the development of segregation in the South had immediate implications both for the embattled defenders and the crusading opponents of racial segregation. If the thesis were sound, then the traditional defenses of segregation were breached and weakened because they pictured the system as entrenched in immemorial and unbroken usage and quite beyond the reach of legal action. And again, if the system were of relatively recent origin and was itself the result of political and legal action, then reformers might take hope that segregation was not all that invulnerable and that the law might be used effectively to bring about change. Usefulness to reformers or embarrassment to conservatives should, of course, never be regarded as an admissible test of the validity of any historical thesis. Yet the time and attention of the author were initially (and for a long time) preoccupied, not with answering historical criticism, but with attempting to correct misreadings, distortions, and misuses of his work by partisans in a raging debate over public policy. No, he had not said that segregation was superficially rooted or easily eradicated; it was rooted in prejudices older than slavery and might

[2] Winthrop D. Jordan, "Modern Tensions and the Origins of American Slavery," *Journal of Southern History*, XXVII (1962), 18–30.

be as difficult to uproot. No, he had never suggested that the period after Reconstruction was a golden age of race relations. Yes, he was quite aware that *de facto* segregation normally antedated *de jure* segregation and that it was widely practiced in several areas before the 1890's. And so forth and so on.

Scholarly response was slower in coming. In publishing his thesis, the author had observed that his subject was relatively new and inadequately explored, that he would inevitably make mistakes, and that he would "expect and hope to be corrected." In this he was not to be disappointed. Initially, however, the corrective scholarship took the form of extending the investigation backward in time and arriving at findings not incompatible with the original thesis. The author had confined his study to the South. In 1961, Leon F. Litwack published a path-breaking study of race relations in the North before the Civil War which demonstrated that Negroes of the free states were segregated "in virtually every phase of existence" and that the Jim Crow system had been thoroughly established in the North before moving South.[3] Subsequent studies of the Northwestern states supported these findings and sustained Tocqueville's observation that "The prejudice of race appears to be stronger in the states that have abolished slavery than in those where it still exists; and nowhere is it so intolerant as in those states where servitude has never been known." [4]

While these findings required no alteration of the thesis as applied to the South, other findings did. Richard C. Wade in 1964 discovered a rudimentary but unmistakable pattern of segregation in some of the larger cities of the antebellum slave states. It was not uniform, never complete, and enforcement was not rigid, but it was undoubtedly present. Since less than 2 percent of the total slave population lived in the cities, however,

[3] Leon F. Litwack, *North of Slavery: The Negro in the Free States, 1790–1860* (Chicago, 1961).

[4] Eugene H. Berwanger, *The Frontier Against Slavery: Western Anti-Negro Prejudice and the Slavery Extension Controversy* (Urbana, 1967); V. Jacque Voegeli, *Free but Not Equal: The Midwest and the Negro During the Civil War* (Chicago, 1967).

and since urban life was so uncharacteristic of the Old South, this evidence did not disturb the validity of the thesis in the rural and typical South. Moreover, Wade's evidence of physical proximity and intimacy of the races in residential, social, and sexual relations in cities — far greater than in the North — gave more emphasis to close contact and association than to separation of black and white.[5] But in South Carolina, according to a study of the Negro during Reconstruction in that state by Joel Williamson published the following year, the slave experience of "constant, physical intimacy between the races" came to an abrupt end with slavery. "Well before the end of Reconstruction," he writes, "separation had crystallized into a comprehensive pattern which, in its essence, remained unaltered until the middle of the twentieth century." In fact, he goes so far as to say that "the pattern of separation was fixed in the minds of the whites almost simultaneously with the emancipation of the Negro." [6] Of course this seemed irreconcilable with the disputed thesis. If Williamson were right about South Carolina, and if that state were not wholly untypical of others, little remained to be said in defense of the thesis.

In the meantime, however, quite opposite conclusions were announced in studies of the other Southern states. A monograph on Virginia by Charles E. Wynes concluded that "the Woodward thesis is essentially sound. Of course certain qualifications must be made, but they do not destroy or greatly impair its essential validity." [7] A book on North Carolina by Frenise A. Logan contained similar endorsement of the thesis and the assurance "that segregation and white supremacy, certainly as they pertain to North Carolina, are products of twentieth-century Southern-White mentality," that "between 1865 and 1898 North Carolina witnessed only a few of 'the policies of proscrip-

[5] Richard C. Wade, *Slavery in the Cities: The South 1820–1860* (New York, 1964).

[6] Joel Williamson, *After Slavery: The Negro in South Carolina During Reconstruction, 1861–1877* (Chapel Hill, 1965), 274, 275, 298.

[7] Charles E. Wynes, *Race Relations in Virginia, 1870–1902* (Charlottesville, Va., 1961), 148–150.

tion, segregation, and disfranchisement' which were later to characterize Negro-White relations in that state." [8] A more recent monograph by Margaret Law Callcott concludes that "Maryland's experience seems to fit readily into this pattern," that the extension of segregation "into public transportation, residential housing, and public accommodations grew out of the purposefully generated racism that marked the return of the Democratic party to power in 1900," and that this was "a new and startling development" with which Maryland Negroes "were frankly unfamiliar." [9] Other works under way or already published supported the thesis in various ways.[10]

In 1968 Joel Williamson compiled and edited briefs of the long adversary procedure up to that date — pro and con, old and new — under the title, *The Origins of Segregation.*[11] One could not hope for fairer or more scrupulous treatment from an editor, nor could one find in the entire roster of contemporary American historians one who combined more naturally a devotion to the pursuit of truth with an adherence to the Shakespearean tradition of adversarial ethic. He placed all parties to the controversy in his debt. That debt would have increased had he been able thereby to put an end to the whole controversy, but unfortunately this lay beyond his power — perhaps beyond the power of anyone. The adversary procedure appears to be endowed with autonomous powers of self-perpetuation.

It surfaced next in the learned journals, in one of which appeared an article elaborating for New Orleans the investigations of Richard Wade on antebellum race relations in Southern cities and coming to much the same conclusions. It was a picture of

[8] Frenise A. Logan, *The Negro in North Carolina, 1876–1894* (Chapel Hill, N.C., 1964), vii–viii, 180.

[9] Margaret Law Callcott, *The Negro in Maryland Politics, 1870–1912* (Baltimore, 1969), ix–x, 134–135.

[10] Vernon L. Wharton, *The Negro in Mississippi, 1865–1890* (Chapel Hill, 1967); George B. Tindall, *The South Carolina Negroes, 1877–1900* (Columbia, S.C., 1952).

[11] Joel Williamson (ed.), *The Origins of Segregation* (New York, 1968).

whites "mixing freely with Negroes in colored taverns, bawdy houses, and dance halls," a scene where Negroes enjoyed "an unusual scope of freedom," where "Negroes, free and slave, made a mockery of the 'Sambo' stereotype," and where "control over the Negro population was, in short, virtually nonexistent." Antebellum segregation laws are presented as evidence of frantic white efforts to curb mixing and separate the races. Unlike Wade, however, this author identifies his picture of antebellum urban experiments in segregation with the twentieth-century system, holding that "both systems effected a thoroughgoing separation of the races . . . in nearly every area of public activity." [12] Unfortunately the author overlooked a previously published study of race relations in the same state which assembled massive evidence that "the Woodward thesis is basically sound for Louisiana between 1877 and 1898. . . . Clearly segregation in Louisiana did not exist before 1898 as a permanent and thorough system of race relations." [13] Called to account for his oversight, the author of the New Orleans study admitted that this city presented "a decidedly unique situation," that his article was "not so much an attack on the 'Woodward thesis' as it was a rejection of certain dogmatic Woodwardian disciples," and that his departure from the former represented only a "delicate difference." [14]

The latest efforts of reconciliation are those of August Meier and Elliott Rudwick, two prominent historians of American race relations. "It is our belief," they write, "that the views of Woodward on the one hand, and Williamson and Wade, on the other, are both correct." [15] The substantive suggestions and findings of

[12] Roger A. Fischer, "Racial Segregation in Ante Bellum New Orleans," *American Historical Review,* LXXIV (February 1969), 926–37.

[13] Henry C. Dethloff and Robert P. Jones, "Race Relations in Louisiana, 1877–1898," *Louisiana History,* IX (Fall, 1968), 322.

[14] Letter to the Editor from Dethloff and Jones and reply from Fischer, *American Historical Review,* LXXV (October 1969), 326–27.

[15] August Meier and Elliott Rudwick, "A Strange Chapter in the Career of 'Jim Crow,'" in a collection edited by the same authors, *The Making of Black America: Essays in Negro Life and History* (2 vols., New York, 1969), II, 15.

these two scholars deserve and will receive full consideration. In passing, however, one is reminded of the efforts of historians at the crest of American nationalism to reconcile North and South on the occasion of the semicentennial anniversary of the Civil War. "Both sides were right!" declared one of them. "Neither could have given in and remained true to itself." [16] One would like to believe that the latest effort of conciliation might be more successful than the earlier one, but the prospect was not very reassuring. In spite of the relatively bloodless and decorous character of the segregation controversy and the civilized deference with which most of the disputants have treated opponents, escalation seemed to be more indicated than abatement — the disposition to "remain true" more prevalent than the impulse to "give in." Investments of time, defenses of vanity, and the loyalties of allies all made their demands. More additions to the already formidable shelf of monographs, pro and con, were on the way, and as the shelf lengthened the patience of readers shortened. One collector of historical fallacies inadvertently contributed one of his own making by impatiently dismissing one side of the argument as "wrong in all its major parts." [17] Even if the preponderance of new studies continued to sustain the disputed thesis, as in the past, they would not resolve the controversy. The prospect of continued proliferation of monographs, pro and con, may well be viewed with dismay.

What is needed is a theory, a model, perhaps a typology of race relations that would conceive of the historical problem of segregation not as one of dating origins at a point in linear time but of accounting for the phenomenon in whatever degree it appears. Such a model might accommodate pieces of evidence that seem contradictory when arrayed in the present formulation of the problem. Several suggestions have been made, to two of which the present writer has been invited to subscribe but finds

[16] Emerson D. Fite, *The Presidential Campaign of 1860* (New York, 1911), 195–96.

[17] David Hackett Fischer, *Historian's Fallacies: Toward a Logic of Historical Thought* (New York, 1970), 148–49.

himself unable to accept. One is "an evolutionary concept of segregation." [18] This suggests a gradual and organic growth over a prolonged span of time, which does not accommodate the evidence of long periods of dormance and sudden bursts of proliferation. A second proposal is that segregation be viewed as "a cyclical development." [19] But this suggests a pattern of regularity and predictability without a causal explanation and without sufficient examples from history to provide a test. A third idea would have it "that industrialization is a master agent of social transformation" in race relations. The most persuasive critique of this theory, however, concludes that "available evidence everywhere sustains the thesis that when introduced into a racially ordered society industrialization conforms to the alignment and code of the racial order." Changes in the racial pattern arise mainly from political, not industrial pressures.[20] No clearer instance of this can be found than in the industrialization of the South. None of the three proposed theories seems very helpful.

More promising are theories derived from comparative history and sociology. Pierre L. van den Berghe works out a typology of race relations based on a comparison of the history of Mexico, Brazil, the United States, and South Africa. The "two ideal types" of race relations that he proposes are the "paternalistic" and its opposite, the "competitive," which he thinks of as parallel with the distinction between *Gemeinschaft* and *Gesellschaft*. The first relies on social distance, the second on physical distance to define status. In the paternalistic system race relations follow a master-servant model, in which a dominant minority rationalizes its rule as a benevolent despotism that regards the subordinates as childish, irresponsible, but lovable in "their place." The subordinate group accommodates to or "accepts" inferior status and sometimes "internalizes" inferiority

[18] Roger A. Fischer, Letter to the Editor, *American Historical Review*, LXXV (October 1969), 327.

[19] Meier and Rudwick, "A Strange Chapter," 19.

[20] Herbert Blumer in Guy Hunter (ed.), *Industrialisation and Race Relations: A Symposium* (London, 1965), 220, 245.

feelings. Role and status are sharply defined on racial lines and social distance is symbolized by elaborate etiquette and "nonreciprocal terms of address." The system accepts miscegenation and concubinage, and the "great degree of social distance allows close symbiosis and even intimacy, without any threat to status inequalities. Consequently, physical segregation is not prominently used as a mechanism of social control and may, in fact, be totally absent between masters and servants living together in a state of 'distant intimacy.' " Racial prejudice often takes the form of "pseudotolerance," with professions of love for the subordinated. Typical relations are "unequal but intimate." In the opposite or *competitive* type of race relations, role and job are no longer sharply defined by race, and class lines begin to cut across race lines as the gap narrows between races in education, occupation, income, and life style. The master-servant model that established social distance with etiquette gives way to sharp competition between the subordinate race and the working class of the dominant race. The latter joins the upper class to form a "*Herrenvolk* democracy" to put the excluded race down. "To the extent that social distance diminishes, physical segregation is introduced as a second line of defense for the preservation of the dominant group's position." Contact declines, miscegenation decreases, racial ghettos appear. Instead of the jolly, irresponsible child-race of paternalistic days, the lower caste of the competitive era appears "uppity," insolent, and aggressive. In place of the condescending tolerance and *noblesse oblige* benevolence of the old days, race hatred and virulent bigotry appear and conflict erupts in lynchings and riots. The shift to the competitive type of race relations is identified by van den Berghe with industrialization, but as we have seen, there are reasons to doubt the necessity of this identification. This author also fails to note anticipations of the competitive type in the paternalistic phase of race relations.[21]

Philip Mason more informally suggests a typology of "three

[21] Pierre L. van den Berghe, *Race and Racism: A Comparative Perspective* (New York, 1967), 25–37.

main stages" based on his personal experience in England, India, and the West Indies, plus a sophisticated reading of Shakespeare's *Othello* and *The Tempest*. Although Mason arrives at his analysis quite independently and includes English class relations as well as colonial race relations, the first two stages of his typology resemble those of van den Berghe. The first he calls "a stage of certainty; a slave is a slave and can be sold, but you can eat with him, talk with him, travel with him." He finds "this same certainty of status and freedom of personal relationship" characteristic of class relations of the old England emerging slowly from the Middle Ages as well as of race relations of old colonial days in India; even in South Africa there was a time when white men could eat with a free black. "The key to this stage," he writes, "is that the relationship is accepted on both sides." The second stage is one of "challenge and rivalry, of growing bitterness, of personal estrangement and aloofness, of insistence on barriers. This means that the top people are frightened." In this stage neither dominance on the one hand nor dependence and acquiescence on the other can be taken for granted, and status is uncertain. The third stage, a "stage of crisis," contains many contradictions and ambiguities and is more hesitantly defined. Here the dominant group, recognizing the inevitable, makes concessions that it finds unwelcome, but may be greeted by "a proud refusal to accept what has long been withheld." Symptoms of the third stage will be familiar enough to observers of the contemporary racial scene in America, but we are not concerned with it here. It does not necessarily follow from the other two, but the first stage "will almost always be succeeded by the second." It is with Mason's first two stages, and especially with their overlapping — as with van den Berghe's two types and their overlapping — that we are mainly concerned in reexamining the disputed origins of segregation in the South.[22]

These and other writers provide numerous examples of race

[22] Philip Mason, *Prospero's Magic: Some Thoughts on Class and Race* (London, 1962), 27–33.

relations in other cultures that illustrate the typologies outlined. We might start, however, with one closer home offered by Philip Mason: "Tom Sawyer and Hucklebery Finn thought it wicked to help a runaway slave to escape but were not at all embarrassed at eating with him; sixty years later, their judgements would have been reversed." In far-off India, he offers a picture of the future Governor of Bombay in 1830 sitting cross-legged on the floor with an old Indian clerk who was instructing him in treasury accounts and whom he addressed by an Indian term of respect meaning "uncle." Then eighty years later the picture of a venerable Brahmin lawyer being pushed rudely from his pony by British subalterns because no Indian was permitted to ride a horse in the presence of his white masters. The explanation, says Mason, is that "so long as there is a formal, and in extreme cases a legal, distinction between the status of persons who consider themselves to belong to different groups — whether they are divided by class or by race — they can often mix on a personal level with easy relaxation. But as the formal nature of the difference diminishes, so the dominant group tends to put up barriers and personal relationships deteriorate." [23]

Both Mason and van den Berghe point out overlappings of their archetypes. Thus the latter writes that "in the process of a society's evolution from a paternalistic to a competitive situation a number of 'survivals' of the paternalistic type can linger on for long periods, even though the society as a whole clearly moves toward a closer approximation to competitive conditions." [24] And in India, Mason observes that "the stage of fixed status and friendly personal relations based on paternalism and dependence lingered on in remote country districts long after it had come to an end in the towns, and in the Indian army never really came to an end until the British went." [25]

The history of race relations in Brazil presents differences from that of Anglo-America and the colonies of Northern Europe, differences that are instructive in ways other than the con-

[23] *Ibid.*, 27. [24] Van den Berghe, *Race and Racism*, 34.
[25] Mason, *Prospero's Magic*, 33.

ventional uses of this popular comparison. The sources of the difference are cultural and arise out of a Catholic society that skipped the Protestant Reformation and much of the Enlightenment. Missing in large measure from its history are the egalitarian, Protestant, achievement-oriented, aggressive-competitive traditions of the Northern societies. In place of the Northern sociological pattern of aggression-withdrawal, Brazil maintained a tradition of dominance-submission. Slaves of Brazil were only the most degraded of several dependent classes that were "embedded in a broad, diversified system of personal domination which pervaded the economy, polity and society of nineteenth-century Brazil." [26]

Florestan Fernandes, the distinguished Brazilian social historian, has spelled out the ironic consequences of this tradition for race relations in his country after the decline and abolition of slavery. Since white men did not entertain "any fear of the probable economic, social, and political consequences of social equality and open competition with Negroes" and therefore felt "no kind of anxiety or restlessness and no sort of intolerance or racial hatred," they erected no "barriers designed to block the vertical mobility of the Negro," no measures "to avert the risks which competition with this racial group might have incurred for the white" — no segregation and no Jim Crow. Instead, they relied tacitly, unconsciously, upon "factors of sociocultural inertia" and "the survival *en masse* of patterns of social behavior that were often archaic," including "the old pattern of race relations, as well as certain social distinctions and prerogatives." This was the old paternalistic heritage of slavery combined with the dominance-submission order of Brazilian society. It was not imposed by threatened lower-class whites, but by the elites, who "tended to maintain rigid, incomprehensible and authoritarian attitudes" toward blacks and "acted as if they still lived in the

[26] I am indebted to unpublished papers of Richard Morse for these insights, especially his "Comments on Professor Carl N. Degler's Paper," read at a meeting of the Organization of American Historians at Philadelphia, April 17, 1969.

past and exaggerated the potential risks of an open liberalization of the social guarantees for Negroes." While the common white "did not feel he had to compete, contend, and struggle against the Negro, the latter tended passively to accept the continuation of old patterns of racial adjustment." What then of the famed "racial democracy" of Brazil? The elites continued to protest benevolence and maintain social distance. They reproved racial prejudice should it appear among new white immigrants, but kept the Negro at arm's length. "He was not openly rejected, but neither was he openly accepted without restrictions," an ambivalence coupling rejection "with apparent acceptance of the exigencies of the new democratic order." This might encourage friendliness and intimacy between races, but as Fernandes says, "The existence, intensity, and closeness of the contact between whites and Negroes were not, in themselves, indisputable evidence of racial equality. All such contact developed within the most thorough, rigid, and unsurmountable racial inequality." Continued protestation of loyalty to democratic ideals and social justice of the new order were combined with preservation of white supremacy. "In the name of a perfect equality of the future," writes Fernandes, "the black man was chained to the invisible fetters of the past, a subhuman existence and a disguised form of perpetual enslavement. . . . And thus was born one of the great myths of our times: the myth of Brazilian racial equality." [27]

Neither Fernandes nor Richard Morse can conceive of anything quite comparable happening in the competitive, aggressive society of the United States.[28] They are certainly right about the Northern or "free" states. Slavery had never struck deep roots in the North, nor had slaves or free blacks concentrated in large numbers, and there had been no soil in which a paternalistic order could really become established. The paternalistic type of race relations existed only rudimentarily and in the main as a

[27] Florestan Fernandes, *The Negro in Brazilian Society* (New York, 1969), 134–37, 164, 178.

[28] *Ibid.*, 135, and Morse, "Comments on Professor Carl Degler's Paper."

fashionable imitation of a romanticized South. Northern slaves were emancipated piecemeal and entered free society as slaves without masters, terribly handicapped but potential competitors of lower class and immigrant whites and threats to their status. Free blacks met immediate measures of segregation and proscription and passed into the rigors of the competitive stage of race relations with little if any cushioning from a paternalistic experience. Race relations in the North and West as described by Leon Litwack, Eugene Berwanger, and Jacque Voegeli fully bear out Tocqueville's observation that the intensity of race feeling was highest in states where slavery had never been established at all. Most of the anti-Negro legislation and rhetoric of the Northwest, however, was inspired not so much by the presence of blacks, which was usually minimal, as by the threat of a black invasion either as bondsmen of an expanding slave society of the South or as a massive migration of black freedmen after the abolition of slavery. It was in this area of the United States that the competitive type of race relations arrived at its first expression and that the sociological model of aggression-withdrawal thrived unchecked by any dependence-submission heritage of paternalism.[29]

The slave society of the South *was* a paternalistic order, perhaps the most paternalistic of the slave societies in the New World.[30] This implies no invidious judgment of comparative benevolence. There is no necessary correlation between benevolence and the patriarchal family, and severity is quite as compatible with paternalism in master-slave relations as in father-child relations. But whatever paternalism implied in race relations, the South might be expected to provide in full measure. The master-servant model was with rare exceptions supreme. Role and status were defined not only by race but by law. Social distance was so wide and so clearly symbolized by punctilious eti-

[29] See footnote 4 above.

[30] This is obviously a disputable position, depending much on one's conception of paternalism. See Eugene D. Genovese, *The World the Slaveholders Made* (New York, 1969), 96, 101, 131, 199.

quette, by terms of address, by gestures and rituals and manners manifesting subservience and dominance that precautions of physical distance, segregation, were rendered superfluous in all but exceptional circumstances. Certainty of status permitted "distant intimacy" and affective bonds between the races and encouraged "pseudotolerance" in the master class. This was what W. E. B. DuBois described "before and directly after the [Civil] war" as "bonds of intimacy, affection, and sometimes blood relationship, between the races," particularly with "domestic servants in the best of the white families." [31]

Even in slavery times, however, not all elements in the Southern population came equally under the ambiance of the patriarchy. Some people simply did not "fit in" conveniently or they responded skeptically or rebelliously to paternalistic authority. These were the free black people, scarcely more than a quarter of a million at most; yet they occupied a disproportionate share of white concern and anxiety in spite of their overwhelming dependency and submissiveness. While nonslaveholding whites generally looked to planters for leadership on the slave issue, there were some who were never happy with upperclass casualness about "distant intimacy" with blacks. And there was an underclass of whites whose social distance from them was not all that secure. These elements tended to concentrate in the antebellum cities of the South. There free blacks and underclass whites were thrown together with slaves under conditions hostile to both the discipline of slavery and certainty of racial status. As Richard Wade observes of the two races in urban setting, "they encountered each other at every corner, they rubbed elbows at every turn," and they lived together on virtually every city block. In most situations, as Wade says, they were "placed in such a way that social distance between the races was maintained even under conditions of close physical proximity." But social distance was not enough to define caste status in the larger cities. This was especially true in New Orleans, where for a time free people of color, largely French-speaking, were almost as numerous as

[31] W. E. B. DuBois, *The Souls of Black Folk* (Chicago, 1903), 183–85.

slaves. Among them were families of wealth and education. Class lines crossed race and caste lines at both ends of the social spectrum. At the lower end, working-class whites competed with and drove Negroes out of many trades, and at both ends of the social scale, the races mixed promiscuously for pleasures and amusements on a more or less clandestine basis.[32]

Under these circumstances, the paternalistic type of race relations began to tip over into the competitive type, and as soon as this happened, physical distance began to supplement or reinforce social distance with various forms of segregation, legal and extralegal. It is true that urban life was rare in the slave South, and that five of the Southern states did not boast a town with as much as 10,000 population. It is also true that the degree and kind of segregation differed from city to city, was rarely rigid and never uniform, and that in no antebellum city of the South did segregation approximate the rigidity and completeness the system was to attain throughout the South in the twentieth century and had already attained in parts of the North by the mid-nineteenth century. Nevertheless it is perfectly clear that segregation in the South made its first appearance in the cities of the slave regime. If the problem is simply that of locating the first appearance of segregation in linear time, this is the answer and there is nothing more to add. But that is not a very significant or challenging problem. More important, it would seem, is the fact that the old paternalistic order of race relations continued to prevail in the rest of the South and that the urban experiments with the competitive order and segregation solutions touched the lives of few Southerners, black or white, during the Old Regime.

The first great crisis[33] in race relations of the South occurred in two parts: first, over the abolition of slavery, and second, over the earlier part of Reconstruction. Both added physical distance between races, but the distances produced by the two ex-

[32] Wade, *Slavery in the Cities,* 55, 277; Fischer, "Racial Segregation in Ante Bellum New Orleans," 926–37.

[33] Certainly greater than the crises over slave rebellions and threats of rebellion in the first third of the nineteenth century.

periences were of different origins and permanency. The first kind was a more or less automatic and simultaneous withdrawal of both master and slave from the enforced and more onerous intimacies and the more burdensome obligations of the old allegiance: duties on the one side, responsibilities on the other. On these, there occurred a general walkout on both sides, and even the master was often disposed to say "good riddance" to this aspect of separation. Bereft of the benefits, he could at least shake off the obligations. The other type of separation sprang from different causes and was generated entirely on the white side. The cause was mainly fright — fear of the known, and real apprehension of a menacing unknown. Not only were the slaves liberated, but many were armed, and there was talk of making them full citizens, giving them the franchise, endowing them with equal political and civil rights, and even distributing the plantation estates among them. This was not only competition, it looked to many whites like a takeover. Racial hegemony itself appeared to be in jeopardy. Temporarily, at least, the immediate postwar years fully qualified for Philip Mason's stage of race relations in which "the top people are frightened." They responded in numerous ways, some of which were to erect barriers and add physical distance between the races. The Black Codes of 1865–66 were mainly concerned with forced labor and police laws to get the freedmen back to the fields and under control.[34] But some of the statutes were segregationist. Three states adopted laws discriminating against Negroes on trains, mainly excluding them from first-class cars. Only one required a Jim Crow car. Public facilities that housed members of both races continued to quarter them separately, most often without statutory requirement as they had before the war, and the new schools followed suit. Such segregation laws and practices as Southern cities had adopted before the war were continued and a few more were added in the immediate postwar years.

[34] Theodore B. Wilson, *The Black Codes of the South* (University, Ala., 1965).

It is in connection with the urban situation that the suggestions of Meier and Rudwick, mentioned above, have particular significance for the disputed thesis about segregation. They have called attention to the segregation of Negroes on streetcars in New Orleans, Richmond, Charleston, Louisville, and Savannah in the early postwar years. In some cities this was a continuation of prewar practices, in some the result of new laws. In each city the black freedmen organized protests, demonstrations, or boycotts against the segregated streetcar system, and in each case the whites backed down and abolished it. This occurred in 1867 in New Orleans, Charleston, and Richmond, and three years later in Savannah and Louisville. This leads Meier and Rudwick to suggest "that because of Negro opposition segregation did indeed decline markedly for a period, at least in public transportation." There were no more Jim Crow cars in those cities until after 1900 and the era of full segregation. "Between these two eras," they write, "southern white practices, at least in the cities under consideration, had been less rigid and less harsh toward the Negroes, and there had been acceptance of integrated streetcars." [35] Two forces were at work here, and neither can be neglected in explaining the history of segregation and race relations in Reconstruction and the period following. There was not only the militancy of the blacks, but also the acquiescence of the dominant Southern whites between the early 1870's and the late 1890's. Neither of these attitudes was to survive that period, and black militancy quickly subsided, but the presence of both at some point is required in accounting for the strange character of race relations while the period lasted.

It has been suggested that economic developments are the main clues to changes in style and type of race relations. The rise of racial antipathy and segregation have been variously associated with commercialization, the growth of the bourgeois, the ascendancy of capitalist and industrialization over precapi-

[35] Meier and Rudwick, "A Strange Chapter in the Career of Jim Crow," 15–19; see also Woodward, *Strange Career of Jim Crow*, 27–28.

talist forms.[36] These correlations doubtless have validity in the long run and in more "normal" circumstances. But circumstances were not "normal" in the postwar South, and race relations did not respond "normally" to economic imperatives, at least not for some time. Slavery collapsed, the old planter regime crumbled, industrialization got underway, and a bourgeois regime of the "New South" took charge. But except for an initial shock and hysteria, race relations responded mainly to political and sociological rather than to economic determinants.

Among the political determinants was a temporarily radical Congress and later radical state administrations that adopted civil rights and antidiscrimination statutes to protect freedmen and for a time showed some disposition toward enforcing them. With their franchise in hand, their brothers in office, and some protection for their civil rights, Negroes were emboldened to try new roles, enter strange places, and experiment with race relations in ways that had been unthinkable before and would become so again.[37] The radical phase of the new freedom passed soon, but while it lasted freedmen were capable of rising to such actions as smashing the incipient Jim Crow streetcar systems in five of the largest Southern cities. But what explains white acquiescence? The presence of radical power helps to explain the initial capitulation, but what explains the degree of white acquiescence until the end of the century? That is more complicated.

For one thing, Southern whites of the ruling class realized very early, usually by 1870, that their initial fright and hysteria were unwarranted. Reconstruction was not going to jeopardize racial hegemony. The North quickly backed away from radicalism. There would be no black takeover, no real competitive stage of race relations, not even a serious disruption of racial hegemony. By and large, the blacks still "knew their place";

[36] Genovese, *The World the Slaveholders Made,* 109–13; van den Berghe, *Race and Racism,* 29–37.

[37] Wynes, *Race Relations in Virginia* (cited in note 7), 68–83; Logan, *Negro in North Carolina* (cited in note 8), 25–41, 180; Callcott, *Negro in Maryland Politics* (cited in note 9), ix, 134–35.

with a few exceptions, mainly political, roles were still defined by race and so was status; social distance prevailed over physical propinquity. The ancient racial etiquette persisted with few breaches, and so did personal relationships, and so did the dominance-submission pattern. There was restiveness among the white working class over competition with free blacks, but that could be contained for the time being. It was true that blacks continued to vote in large numbers and to hold minor offices and a few seats in Congress, but this could be turned to account by the conservative white rulers who had trouble with white lower-class rebellion. Black votes could be used to overcome white working-class majorities, and upper-class white protection was needed by blacks under threat of white lower-class aggression. Many reciprocal accommodations between upper-class whites and blacks were possible under the paternalistic order.[38]

For all the differences between the two countries in the history of their racial relations, one is reminded of the South of this era in Roger Bastide's description of Brazil in the postabolition period:

> Slavery had disappeared, but the consequence was not that the mass of Negroes rose into the global community; the Negroes remained just where they were before, not forming a competitive group. It is precisely because they did not constitute a danger to the traditional social structure, because they did not threaten the whites' status, that the latter did not feel fear, resentment or frustration toward colored people. Personal, emotional relationships could thus come about between whites and blacks. . . .
>
> Paternalism prevented tensions and softened the relations between races. But, at the same time, it . . . institutionalized the subordination of the Negroes, who could only benefit from the protection of the whites, or from a certain familiarity in the whites' treatment of them, on condition that they 'knew their place' and proved their deference, gratitude and respect. It was therefore an instrument of political and economic control, which, by avoiding the competitive relations . . . by preventing a struggle, and by rendering useless

38 C. Vann Woodward, *Origins of the New South, 1877–1912* (Baton Rouge, 1951), 103–106.

any wish for collective mobility on the part of the Negroes, assured supremacy and security to the white class. Under these circumstances, one can understand why prejudices are at a minimum in a paternalist society, or, at least, why they remain latent rather than finding external expression. The reason is that they are unnecessary.[39]

Of course there was more white racism and incipient segregation in Southern paternalism, both during slavery and after, than in Brazil, and the South did not make the distinctions in shades of color that Brazilians made. Among other differences, the Brazilian Negro did not enjoy even such experience of political power and participation as their Southern cousins precariously enjoyed during and after Reconstruction. Even so, a good deal of similarity exists between the Southern and the Brazilian extension of white paternalism under bourgeois auspices after the end of slavery. There is also similarity in the bourgeois curtailment of paternalistic benefits under the two regimes. The Southern extension ended earlier, but it served many of the same purposes.

The distinctiveness of the Southern style required the heritage of Reconstruction as well as the heritage of slavery, the political as well as the sociological ingredients to subordinate for a time the economic imperatives of the competitive order in race relations. While the radical promise of black equality and the threat to white racial hegemony quickly faded, the black franchise lingered on for thirty years, and the civil rights statutes remained on the books — honored mainly in the breach though they were. Black members, reduced in number but still a conspicuous presence, continued to sit in Southern state legislatures, and black voters sent as many congressmen to Washington in the last quarter of the century as they had during Reconstruction. They could not be ignored politically as they could be after disfranchisement, and they had something to offer as well as something to beg of their paternalistic patrons. Together the oddly paired combination held at bay the worst of the fanatics and for a while

[39] Roger Bastide, "The Development of Race Relations in Brazil," in Hunter (ed.), *Industrialisation and Race Relations* (cited in note 20), 14–15.

stemmed the tide of proscription, segregation, and disfranchisement. In the patrician style, white conservatives sometimes denounced Jim Crow laws as "unnecessary and uncalled for" and "a needless affront to our respectable and well-behaved colored people." They ridiculed Jim Crow measures as preposterous and snobbishly identified the demand for them as lower class. One declared that he would rather "travel with respectable and well-behaved colored people than with unmannerly and ruffianly white men," and that if a Jim Crow car were needed it was for the latter. They made concessions, but when the first state Jim Crow law for trains was passed in 1887, a conservative paper rather shamefacedly admitted it was done "to please the crackers."

In the years 1900–1906 the Jim Crow movement swept the cities of the South, and one city after another passed municipal ordinances or applied new state laws requiring segregation of the streetcars throughout the former Confederacy. Messrs. Meier and Rudwick, who investigated the Negro protest that ended the abortive Jim Crow streetcar system in the 1860's have also done a revealing study of its recrudescence in the 1900's. "For Negroes the new order was startling, even shocking," they report. The Negroes' response was an attempt to repeat the triumph of the 1860's. Between 1900 and 1906, they staged boycotts lasting from a few weeks to two or three years in more than twenty-five Southern cities. The protests were "almost entirely led by conservative business and professional men," who were "known as impeccably respectable men rather than as radicals or firebrands," and the movement is described as "conservative in the sense that it was seeking to preserve the status quo — to prevent a change from an older and well-established pattern." Black protest leaders naturally looked to their conservative white allies for support. "Universally the effect [of the boycotts] was startling to the white population." A few conservatives responded with appeals to maintenance of the "traditional harmony," and respect for "old Mammy," but the needed support was not forthcoming. After a few temporary successes, "in the end, the boycott movements against Jim Crow trolleys failed in all of the cities where they were initiated."[40]

The paternalistic order under bourgeois custody had been insecure for some years, and by this time the collapse of the "traditional harmony" and the political alliance that sustained it was almost complete. Its collapse was but one of several reasons for the rise of as harsh, as rigid, and as brutal and universally segregated a system as any example of the competitive order of race relations on record, surpassing that of the North in completeness and in the opinion of a South African white in 1915, as thorough as that of his native land. The forces that combined to bring about the change have been sketched elsewhere by the writer and there is no reason for repeating them here.[41] They are complex and numerous and some of them, important ones, were neither regional nor national in origin. One factor requires further comment, however, in view of the earlier reference to typologies and models of race relations. "Models" are always suspect by historians, for they know that however suggestive models and typologies may be for hypotheses, concrete history never fits into them. Historians have somewhat the same reservations about comparative history. One model suggested that the tipping point between the paternalistic and the competitive order of race relations came when "the top people are frightened." The top people of the South *were* in this instance frightened to some degree, all right, but they were frightened by the white lower class, not by the blacks. Some of them may have sounded like it at times, but they were usually "talking for the crackers." The ratio of whites to blacks has always set the South off as distinctive from other societies of "Plantation America," and not enough has been made of this white preponderance in comparing slavery systems and race relations. A great proportion of Southern whites had never been slaveholders or come personally within the paternalist system. They had been

[40] August Meier and Elliott Rudwick, "The Boycott Movement Against Jim Crow Streetcars in the South, 1900–1906," *Journal of American History*, LV (March 1969), 756–75.

[41] Woodward, *Strange Career of Jim Crow*, 69–109.

economically distressed and politically rebellious since Reconstruction. The pinch of depression in the 1890's undoubtedly intensified their unhappiness, but no momentous shift in the means of production or its ownership intervened to account for the rush into a new era of race relations, "competitive" or otherwise. The upper class desertion of the blacks and capitulation to racist demagogues and their ruthless program of proscription, disfranchisement, and segregation stemmed from political not economic pressures. The new order of race relations was shaped and defined by political means and measures, and they came not to meet the needs of commerce or industry but the needs of politicians.

For all that, the Manichean categories of the "paternalistic" versus "competitive" typology do not do full justice to historical realities. It remains to admit that segregation of itself might be regarded as in some measure a modification or extension rather than an end of the paternalistic order. The escalation of lynching, disfranchisement, and proscription reflected concessions to the white lower class. So did legal segregation. But while segregation diminished the range of "distant intimacy" and "pseudo-tolerance," it was not inconsistent with a modified paternalism of the upper class whites.[42] It provided a legal definition of role and status that formalized social distance sufficiently for the older personal and paternalistic black-white relationships to be carried on. Only in this way can the differences that were still observable between Southern and Northern styles of racial relations in the mid-twentieth century be accounted for. Viewed in this way the first half of the twentieth century may be seen as an extension of the overlap between paternalistic and competitive systems. Not until the Civil Rights Movement knocked out the institutional foundations of the modified paternalism, already undermined by competitive encroachments, did the South finally join the North in the fully competitive phase.

To return finally to the question of adversary procedure, the

[42] I am unable, however, to go so far as Guion Griffis Johnson goes in identifying segregation with paternalism. See her article, "Southern Paternalism toward Negroes after Emancipation," *Journal of Southern History*, XXIII (1957), 483–509.

writer has confessions of fault and responsibility bred of further thought. While he did take pains to say in so many words when he originally advanced his thesis that the "new era of race relations was really a heritage of slavery times," [43] he probably did not sufficiently emphasize the paternalistic character of those relations. In neglecting to do this, he (unconsciously?) permitted the hopeful but unwary modern reader to identify such casualness and permissiveness as graced that remote Southern interlude of paternalism with the type of open, color-blind egalitarianism to which the modern liberal aspires. That was a mistake. It was also a mistake not to warn the unsuspecting that although the permissiveness of that era allowed such precocious Southern liberals as George W. Cable and Lewis H. Blair to speak their minds on race relations, they did not speak for the South of their day, but against it. And in his sympathy for the Populists, he probably neglected to point out that their egalitarian approach to Negro voters was itself mixed with a good deal of paternalism. Sensing these shortcomings without precisely spotting them, but realizing something was wrong, critics were lured into an adversary posture. If it would help to sidetrack the controversy, the writer would gladly concede that paternalism was never a way out for the Negro. It helped to temper race relations and postpone the worst rigors of segregation, not so much as Fernandes and Bastide describe it doing in Brazil, but with much the same effect and at the same cost. It bound the black man to the past, to dependence and withdrawal, to "his place." It avoided much competition and some overt tension, but it closed the door to independence and the future.

Where does this leave the adversary procedure? If it leaves the writer as advocate for the plaintiff, he would enter a plea of *nolle prosequi*. If spokesman for the defendant, a plea of *nolo contendere*. If neither plea be granted and the procedure be involuntarily protracted, he will persist as adversary only in conformity with the Shakespearean admonition.

[43] Woodward, *Strange Career of Jim Crow*, 43.

10

The Elusive Mind of the South

TIME has dealt gently and critics generously with W. J. Cash and his book, *The Mind of the South.* In their searching reassessments, the rebellious young have left his reputation untouched. Their elders have been generally protective of him, as of one of their own. Being a nonacademic and a nonprofessional historian, he has been spared the rivalries and exempted from the standards of the schools and the professionals. His license as a free lance helped him earn a charmed immunity. Scholars often quote him unjealously and flatteringly with a freedom they normally begrudge their fellow academicians. His early and tragic death by suicide following so closely on the publication of his book probably discouraged rigorous reappraisals. There were only two hostile reviews and the author of one of them, Donald Davidson, publicly lamented that the news of Cash's death came too late to enable him to cancel publication of the review.[1]

There are, of course, more positive and deserving reasons for the reputation the book has enjoyed. No reader of any percep-

[1] Donald Davidson, "Mr. Cash and the Proto-Dorian South," *Southern Review,* VII (1941–1942), 1–20. Twenty years after the book was published Dewey Grantham, "Mr. Cash Writes a Book," *The Progressive,* XXX (1961), took some exception to Cash's interpretation in an otherwise favorable tribute; ten years later Eugene D. Genovese, *The World the Slaveholders Made* (New York, 1970), 137–50, took much more serious exception.

tion can fail to sense the passionate involvement of the author in his subject, nor fail to be torn by the love-hate intensity of his feeling for the South. His studied lightness and occasional flippancy do not conceal a dedicated concern, a personal anguish. Here, obviously, was a man writing his heart out about the subject that was dearest to him. If he failed, it was surely not for want of trying or caring. No one would go so far as to pronounce a flat verdict of failure upon a work of such grace and originality, such haunting cadences and gifts of phrase making. They stick in the memory, those phrases — the "Proto-Dorians," the "hell-of-a-fellow complex," the "savage ideal," the "lily-pure maid of Astalot" and "the ranks of the Confederacy rolling into battle in the misty conviction that it was wholly for her that they fought." We are all grateful for these phrases, and our common stock of insights and perceptions is surely the richer for some of his contributions.

With these assets, the book made its way steadily upward in prestige during the forties, fifties, and sixties. The loyalties of several generations of college students, grateful for temporary relief from textbook prose and pedantry, have been enlisted. Southerners have had to concede from the start the authenticity of Cash's credentials as a member of the club. Non-Southerners have been reassured by his offhanded dismissal of regional pieties, myths, and romances, as well as his confirmation of some of their worst suspicions. His *Mind* was considered by some the perfect foil for the contemporaneous novel *Gone with the Wind*. Liberals have been won over and gratified by Cash's uncompromising attacks on lynching and other varieties of inhumanity, injustice, oppression, and brutality, as well as by his criticism of the ruling class, his friendliness to labor, and his benevolent attitude toward the Negro. Toward the South as a whole he strained to achieve balance and fairness. In eloquent passages that concluded his book, he held up two images:

Proud, brave, honorable by its lights, courteous, personally generous, loyal, swift to act, often too swift, but signally effective, sometimes terrible in its action — such was the South at its best. And such at its best it remains today.

And then by contrast:

> Violence, intolerance, aversion and suspicion toward new ideas, an incapability for analysis, an inclination to act from feeling rather than from thought . . . attachment to fictions and false values, above all too great attachment to racial values and a tendency to justify cruelty and injustice.[2]

And such remained the South's characteristic vices.

The bold simplifications, the memorable formulas, the striking symbols, and the felicitous phraseology of the book have fixed it firmly in the esteem of journalists and popular writers. Their unstinted praise would gratify the vanity of any author. The book is quoted, paraphrased, and plagiarized so regularly as to have practically entered the public domain. Nor have journalists been its only champions. Social scientists, especially sociologists, seem to have a special affinity for the book. Some historians, even historians of the South, have praised and recommended *The Mind of the South*. The original reviews in the learned journals were generally favorable, and the numerous references to Cash in *Writing Southern History* are invariably respectful or laudatory.[3] Those historians who have had reservations and serious criticisms have thus far kept their counsel. They therefore share responsibility for permitting a book that has for so long a time virtually escaped serious professional criticism to become established in the remarkable prestige this work enjoys. (As one of the original reviewers, I assume my share of the responsibility.[4]) It would be impossible to prove, but I would venture to guess that no other book on Southern history rivals Cash's in influence among laymen and few among professional historians.

[2] All quotations from Cash, unless otherwise noted, are from the final edition of *The Mind of the South* (New York, 1941).

[3] Arthur S. Link and Rembert W. Patrick, eds., *Writing Southern History* (Baton Rouge, 1965), 150, 350, 375, 384, 442.

[4] *Journal of Southern History*, VII (1941), 400–401. A few "reservations" were noted, however.

Of all the Southern historians of his generation, or indeed of recent generations, W. J. Cash is the only one so far to have become the subject of a formal biography, a very sympathetic one by Professor Joseph L. Morrison.[5] With the subtitle, *Southern Prophet,* this work does not undertake a critical analysis of Cash's book. It simply passes on approvingly what the author would seem justified in assuming to be the general consensus, even in academic ranks, that Cash's *Mind* represents "original historical writing of the highest order." It would now seem about time for a critical reappraisal of that consensus and of the book that inspired it.

One of the rules of criticism, more respected in the breach than in the observance, is that a writer should not be criticized for what he did not undertake to do. I shall do my best to observe the rule, but Cash does not make this easy by presenting us with a work entitled *The Mind of the South* based on the hypothesis that the South has no mind. If his assumption is valid, he could not very well be expected to undertake a history of something that did not exist, and the critic is therefore doubly restrained from criticizing him for his neglect. He is perfectly serious about his hypothesis. He cites more than once and with his endorsement Henry Adams's patronizing quip from the *Education* that "Strictly, the Southerner had no mind; he had temperament." Approving that dubious perception that Adams based on the few Virginians he knew as fellow students at Harvard, Cash goes on to elaborate:

> From first to last, and whether he was a Virginian or a *nouveau,* he [the Southerner] did not (typically speaking) think; he felt; and discharging his feelings immediately, he developed no need or desire for intellectual culture in its own right — none, at least, powerful enough to drive him past his taboos to its actual achievement.

Certainly it would be unjust to criticize the author for not undertaking an intellectual history of the South, and no such

[5] Joseph L. Morrison, *W. J. Cash: Southern Prophet, A Biography and Reader* (New York, 1967).

criticism is intended. But if he were convinced that the Southerner has a "temperament" but no "mind," that he "felt" but did not "think," he might have more accurately entitled his book, "The Temperament of the South," "The Feelings of the South," or more literally, "The Mindlessness of the South." One might dismiss this as quibbling and quote the old chestnut of a metaphysical wit, "No mind, never matter." But it does, I am afraid, matter. This book was published just two years after the appearance of Perry Miller's precedent-shattering book, *The New England Mind.*[6] It is true that Cash used his title earlier for an essay in the *American Mercury.*[7] But Miller's formidable example might seem to have discouraged Menckenian buffoonery with such words. And of course Vernon Louis Parrington, without the depth and sophistication of a Miller, had originated "The Mind of the South" as a title two years before Cash adopted it for his essay.[8] It was not, then, as if Cash was wholly unaware of the responsibilities such a title entailed.

It is one thing, however, to take such liberties, and quite another to justify unloading the responsibilities entailed by denying their existence. Not only does Cash maintain that the Southerner had no mind, but also that "he developed no need or desire for intellectual culture," and in fact achieved none.[9] One is thus presumably prepared for the omission of reference to Southern minds in a book on the Southern mind. Jefferson is mentioned three times, but only in passing and only symbolically, Calhoun twice and Madison once in the same fashion, John Taylor of Caroline and John Randolph of Roanoke not at all. Nor is there any mention whatever of William Byrd, John Marshall, George

[6] Perry Miller, *The New England Mind: The Seventeenth Century* (New York, 1967).

[7] W. J. Cash, "The Mind of the South," *American Mercury* (October 1929), reprinted in Morrison, *Cash,* 182–98.

[8] Vernon Louis Parrington, *Main Currents in American Thought,* II (New York, 1927), 3–179.

[9] Cash would have been on firmer ground had he maintained that the arts and sciences were neglected because of the diversion of creative energies into the defense of slavery.

Fitzhugh, Edmund Ruffin, Hugh Swinton Legré, or Alexander H. Stephens — and a long list of comparable worthies. Perhaps Cash was simply not interested in such minds — and again his freedom to write about what he wishes is conceded, and likewise the rule against criticizing him for doing so. But he goes further to contend that the South had no intellectual achievement, or none worth mentioning, or that such minds as it produced were "the exceptions that prove the rule" — as if all intellectual achievement were not exceptional. The reason offered is that such achievement "never reached any notable development save in towns, and usually in great towns." Presumably contrary evidence from Jefferson to Faulkner are also "exceptions that prove the rule."

But further than that, he even contends that Southerners had "no need or desire" for such endeavors. This in the face of one of the most urgent needs and agonizing desires of any society in the Western world. There were the crying needs for reconciling their bourgeois origins with their antibourgeois professions and institutions, their liberal birthright with their reactionary yearnings, their revolutionary premises with their conservative conclusions, their Enlightenment optimism with their romantic pessimism, John Locke and Adam Smith with Robert Filmer and Walter Scott, egalitarianism with slavery and patriarchy, doctrinaire constitutionalism and love of union with nullification and secessionism. Compared with the issues separating Cotton Mather from Ralph Waldo Emerson, the distance between Thomas Jefferson and George Fitzhugh must be measured in light years. Louis Hartz has called it the "Reactionary Enlightenment" and pronounced it "one of the great and creative episodes in the history of American thought," or more extravagantly, "the great imaginative moment in American political thought, the moment when America almost got out of itself." [10] This may well be an exaggerated estimate. Certainly Mr. Cash has no obligation to agree, and the critic has no right to attack him for writing about something else. But the critic does have

[10] See above, pp. 108–110.

the right and the obligation to protest the blindness and insensitivity which can deny the very existence of obvious problems of the mind, urgently felt needs for solutions, and agonized and fantastically ingenious, if unsuccessful, strivings to solve them. For this insensitivity, I submit, Cash stands indicted.

It may be contended that Cash was really addressing himself to more abstruse and difficult problems than those of the mind — namely the ethos of a people, the prevalent tone of sentiment, the essential temperament that distinguishes their style of life, the rhythm of their responses, the very character of their history. These *are* indeed difficult problems, and they are valid subjects for historical investigation — whether they may properly be called "mind" or not. Assuming these problems to be his real subject, what can be said of his history of the South? Is it comprehensive and balanced? Does it give due attention to all classes, aspects, areas, and periods? The answer on all counts is clearly "No." The author, in fact, is highly selective in the attention he gives to various classes, areas, and periods, neglecting some, emphasizing others. Again, granting the author the freedom to write about what he wants to, it is well to be warned of what he is slighting or omitting, since he gives no notice himself, and his title is not very helpful.

In the first place, this is a book about the "mind" (temperament, sentiment, character, myth) of the Southern whites. This is not to suggest that Cash is blind to the importance of the Negro in Southern history. Far from it. On the contrary, the Negro is of central concern. The black presence is, in fact, a major component of the white mind. One of the longest entries in the index of his book is on the Negro. Furthermore, Cash is outspoken about the wrongs, injustices, discriminations, exploitations and brutalities suffered by Negroes. No one has written with more indignation about lynching. No one is more alert to the impact the blacks had upon the white family structure, sexual attitudes, religious practices, white violence, politics, and myth. "Negro entered into White man," he writes, "as profoundly as White man entered into Negro — subtly influencing every gesture, every word, every emotion and idea, every attitude." The Negro, in fact, is amply treated in so far as he is an influence, a cause, an effect, a victim, a phobia, a myth of white behavior, institutions, and history. But we are told very little about the Negro's own "mind," temperament, emotions, myths, and attitudes. This is not because he feels disqualified for reason of his race. He does not hesitate to tell us that the Negro is "one of the world's greatest romantics and one of the world's greatest hedonists" — that he has a great weakness for rhetoric, and that all these traits had their impact on white behavior, but it is primarily for their influence that they are noted, not their intrinsic significance or origins. To have assumed that responsibility would have required a great deal more attention to the institution of slavery and the slaves themselves than the few pages devoted to the subject.[11] The neglect of slavery and slave mentality is almost as striking as the neglect of the black mind as a component part of the "mind" of the South.

If Cash plays favorites among races and institutions, he also has his preferences among periods. He does say in so many words that "the two hundred years since Jamestown must not be forgotten," but the fact is he does forget them very quickly and brushes over them hastily. One would have thought that the historian of the mind, mind of any description, would have had more interest in origins, especially in the first two centuries. But by page nine we are down to Eli Whitney, and by page twelve the cotton boom is in full stride. This is really a book about the century of Southern history from the 1830's through the 1930's, and considerably more than half of it is devoted to the twentieth century. Most of the few pages that are begrudged the colonial period are given over to belaboring the defunct Cavalier myth. He does admit that a genuine aristocracy arose in this "narrow

[11] "The growing of cotton," wrote Cash, "involves only two or three months of labor a year, so even the slaves spent most of their lives on their backsides, as their progeny do to this day." Quoted in Morrison, *Cash,* 183.

world," this "relatively negligible fraction," of tidewater Virginia and bits of South Carolina and Louisiana. As we shall see, Cash had a peculiar attitude about Virginia, and it was essential to his thesis to belittle the significance of aristocracy. He concedes that small numbers of aristocrats settled widely over the South, but dismisses their importance on the ground that "the total number of families . . . who were rationally to be reckoned as proper aristocrats came to less than five hundred — and maybe not more than half that figure." (And how many families does it take to make a proper aristocracy?) It would seem that influence and models were more important than numbers to a historian of the mind. "But this Virginia," he writes, "was not the great South." And again, "Prior to the close of the Revolutionary period the great South, as such, has little history."

What was this "great South" of Cash's history? Granted the Tidewater was physically a small part, did the "great South" not include the Bluegrass country, the Delta, the Gulf Coast, the Ozarks, the whole great trans-Mississippi South? If so, there is little indication of the fact in his history and in the attention he gives these vast subregions. Natives of Arkansas, Louisiana, and Texas will not find their states in the index of the *Mind,* nor will natives of many other states, with the notable exception of North Carolina. Nor will they find much reference to their states in the text. A map of Cash's great South drawn to scale according to his span of attention and interest after the fashion of *The New Yorker's* Map of the United States, would be as strange a specimen of cartography. The subregion of the Appalachian foothills stretching from Gaffney, South Carolina, Cash's birthplace, through Charlotte, North Carolina, his home as an adult, and tapering northeast to the Potomac would loom larger than all the rest, and North Carolina would easily qualify as the New York of the great South. "Southern industry" for Cash is the cotton mills, and "Southern labor" the lint-heads who work them. When Cash writes in his lyrical apostrophe to nature in the South of "the booming of the wind in the pines," we know that those pines were rooted in the red soil of the Carolina hill country.

A great deal of historical relativism is at work here, the relativism of time and place, of class and race, of period and environment. Cash's *Mind* was an outgrowth of the Great Depression, the "Great Blight," he calls it. The hard-bitten thirties were a time of reckoning, the moment of truth for pretenses of all sorts. Myths went into bankruptcy as often as banks. Cash's hill country was as hard hit as any part of the South, but it had fewer grandiose pretenses. It had never been a land of great plantations, many slaves, or much in the way of aristocracy and fancy living. In the harsh depression climate of the thirties, this subregion of plain folk and small-town industries came belatedly to flower. Thomas Wolfe of Asheville was its poet, Paul Green of Chapel Hill its playwright, James Agee of Knoxville its reporter, and Jack Cash of Charlotte its historian. They felt it was at last their turn to speak for the South, the South as *they* knew it. They had put up long enough with the pretensions and poses of remote and high-living parts and their so-called aristocrats. It was time for an unbiased history of the South from the hill-country point of view. Chapel Hill was headquarters of the renaissance, and Cash identified with its down-to-earth school of sociologists and the "ancient feud" of the Tar Heels with the "Mountains of Conceit" on either side, as well as the new feud with the Nashville Agrarians to the west. Virginia, Cash thought, had had an "artificializing influence." The yeoman farmers were "probably the best people the South has ever produced in any numbers, and its chief hope today." Even the poor whites were "definitely superior, in respect of manner, to their peers in the rest of the country." Posing the question, "What then is our Southern tradition?" he replied: "The best way to answer that, I believe, is to remember who we were and are. . . . The answer to that is that we were plain people in general in our origins." [12]

[12] *Ibid.*, 297.

Recalling our concession of an author's right to choose his subject and the rule that restrains us from criticizing him for not writing on something else, we will call all the preceding discourse "noncriticism." We will call it an effort to discover what Cash *is* writing about. If any note of criticism has inadvertently crept into this description, it is to be deplored as an infringement of the "rule." But now that we have discovered what he is *not* writing about, we may proceed unencumbered by the rule to criticize what he *is* writing about.

Cash is writing about many themes — themes of romanticism and hedonism, individualism and social irresponsibility, complacency and sentimentality, the ethic of leisure, the disposition to violence, the weakness for rhetoric, and "the savage ideal." There is substance and validity in all of these themes. I have no desire to deny the variety and rich texture of his thought, nor is this the place to consider these themes individually. I think it is fair to say, however, that he has skillfully interwoven all these individual themes into two fundamental theses that provide the warp and woof of his book. These are *the thesis of unity* and *the thesis of continuity* — the fundamental unity of the Southern mind and people, a spiritually solid South, and the continuity of Southern history, at least since the American Revolution. On the soundness of these two theses I believe the integrity of his book depends.

Before proceeding to examine the validity of these two theses as Cash applies them in interpreting the mind of the South, it is important first to agree on certain basic assumptions. It is taken for granted that some degree of unity and some measure of continuity are inherent in the subject. Otherwise it could not be intelligently treated by a historian or properly be said to have a history. There must have been at least enough unity in the South to give its people a consciousness of being Southern — whatever else they may have been in addition, including being Americans. This implies a common history, a collective experience distinctive enough to have imparted shared cultural values and symbols of identity. All these things the South undoubtedly had. This kind of unity also implies a certain amount of continuity, enough at least to make it possible to transmit a culture from one generation to the next and from one period to the next. This the South also had, whatever the breaks that occurred in the continuity and however sharply they were felt. So much we may take as given. What is at issue is not the existence, but rather the character and degree of unity and continuity. These are relative and not absolute questions.

In his opening sentence Cash proclaims the existence of "a profound conviction," which he obviously shares, that the South is not only "sharply differentiated from the rest of the American nation," but that it exhibits "within itself a remarkable homogeneity." He quickly concedes that there is "an enormous diversity" as well, but this is overshadowed by the unity of "one South":

> That is to say, it is easy to trace throughout the region . . . a fairly definite mental pattern, associated with a fairly definite social pattern — a complex of established relationships and habits of thought, sentiments, prejudices, standards and values, and associations of ideas, which, if not common strictly to every group of white people in the South, is still common in one appreciable measure or another, and some part or another, to all but relatively negligible ones.

Cash says his generalizations of unity concern "white people in the South," but when the Negro is mentioned it is to emphasize how his traits reinforced and resembled those of whites. Any tendency to conflict, whether of class or economic interest or political doctrine or religious dogma, is played down or dismissed as unimportant compared with the over-all consensus. Writing of the "common white" he says: "Add up his blindness to his real interests, his lack of class feeling and of social and economic forces, and you arrive, with the precision of a formula in mathematics, at the solid South." This formula explains "how farmer and white-trash were welded into an extraordinary and positive unity of passion and positive unity of purpose with the planter," and how "when the guns spoke at Sumter, the masses sprang to arms, with the famous hunting yell soaring in their throats. . . ."

At first glance it would seem plausible to claim for Cash the distinction of anticipating the consensus school of American historians by more than a decade. But this would be misleading, for it underestimates his genuine insight into class relationships in the South. It is not that he is unaware of objective conflict of class interests in Southern history. His pictures of the plight of white-trash, croppers, and lint-heads bear evidence of such awareness. His point about the Proto-Dorian consensus is that it flourished in spite of conflict. The formula for his Solid South, as he says, was the common white's "blindness to his real interests, his lack of class feeling." The lower-class white's obsessive anxiety for racial hegemony trapped him into submission to upper-class hegemony. More than it did anywhere else in the country, this class hegemony prevailed in the South, and it survived in various guises the real breaks in historical continuity. This insight was Cash's main contribution and he deserves full credit for it. And in so far as the Proto-Dorian consensus may be said to have unified the South, and in so far as the consensus survived the discontinuities of the South's history, the thesis of unity and continuity are sustained.

The trouble is that the emphasis on consensus, on "one South," the Solid South, ignores a great deal of evidence of disunity and dissent that does not fit his thesis. One category of dissent swept aside is antislavery sentiment, not only that of Virginians of the Revolutionary generation and the debates of 1831–32, but that which persisted thereafter in the abolitionist efforts in Tennessee in 1834 and Kentucky in 1849. Neglected also are Southerners who became antislavery leaders, such as James G. Birney and his young friends William T. Allen of Alabama and James A. Thome of Kentucky, the Grimké sisters of Charleston, David R. Goodloe of North Carolina, Cassius M.

Clay, John G. Fee, John Rankin, Samuel Crothers, and the Dickey brothers, William and James, of Kentucky, as well as Moncure Daniel Conway of Virginia. It is true that many of these antislavery leaders left the South, but the sentiment they represented was strong enough to support the view of Kenneth Stampp that "the contention of planter politicians that the South had achieved social and political unity appears . . . to have been the sheerest of wishful thinking." [13]

Further evidence of the wishful thinking of the politicians and departures from Cash's thesis of consensus lies in the great body of Southern Unionist dissenters during Secession and Civil War. During the war they furnished some 200,000 troops to the Union Army, and afterward more than 22,000 of them, scattered fairly evenly among the Confederate states, risked the hostility of neighbors and the stiff costs of courts to file claims for more than $60,000,000 for supplies furnished the Union Army.[14] Another body of Southerners that confutes the consensus are the native Republicans, whites as well as blacks. The bulk of the former were of the common folk, but they included former planters, slaveholders, and businessmen of wealth. The post-Reconstruction dissenters, independent farmer-labor and green-back parties such as the Readjusters who took over Virginia from the Redeemers, are apparently unobserved. The white consensus on race policy is exaggerated, and significant differences over this subject between conservatives, liberals, radicals, and fanatics go unmentioned. Cash's effort to fit the Populist upheaval into the consensus had best be passed over in charitable silence.

On the fluent rhetoric of Proto-Dorian unity the "mind" of the South is thus pictured as sliding undisturbed over the divisive crises of Secession, Civil War, Emancipation, Reconstruction, Redemption, and Populism, emerging as undivided and un-

[13] Kenneth M. Stampp, "The Fate of the Southern Antislavery Movement," *Journal of Negro History*, XXVIII (1948), 22.

[14] Frank W. Klingberg, *The Southern Claims Commission: A Study in Unionism* (Berkeley, 1955), 17–19, 101, 157, 207, 209.

shaken as ever. And so also with subsequent upheavals such as the Socialist movement and Hobsbawm-type "primitive rebels" in Texas and Oklahoma and the peace movement that broke out in several states during World War I.

The second major thesis, that of the continuity of Southern history, is really an extension of the thesis of unity to the dimension of time. Not only were the Southern people undivided, but their history from first to last has been undivided by significant breaks and has flowed with essentially unbroken continuity from its sources to the present. Unbroken continuity has been persuasively suggested as a characteristic that accounts for the uniqueness of American national history. With the oldest constitution of any nation, with political parties, executive office, legislative bodies, and judiciary, along with basic institutions and traditions dating continuously from eighteenth-century origins, the United States does indeed enjoy a history of uniquely unbroken continuity.[15] The history of the South, on the other hand, would seem to be characterized more by *dis*continuity, one trait that helps account for the distinctiveness of the South and its history.

Academic historians, it should be acknowledged, are prone to exaggerate the importance of their periods, eras, and epochs, the changes wrought by the beginnings and ends of them, and the breaks they are supposed to cause in the flow of history. Periodizations are both a necessity and a convenience in the writing and teaching of history, but the historian often comes to have a vested interest in their significance and tends to overemphasize such discontinuities as they represent and neglects the continuities that underlie them. At stake are the prerogatives of specialists and the justification of required courses. Academics have resisted Maitland's idea that history is a "seamless web" and rejected "holistic" doctrines generally. They would do well to heed any fresh ideas from outside their ranks that challenge their preconceptions. Cash's emphasis on continuity is such a challenge. His insistence on the durability of the folk mind, the cul-

[15] Daniel J. Boorstin, *The Genius of American Politics* (New York, 1953).

tural heritage, and the anthropological constants in Southern history is a healthy corrective. The contribution he made by demonstrating the persistence of the Proto-Dorian consensus and upper-class hegemony has been pointed out. All this lends support to his thesis of continuity.

Granting all this, we must remember that the issue of continuity is a relative one and that the history of America apart from the South offers instructive contrast. Among the major monuments of broken continuity in the South are slavery and secession, independence and defeat, emancipation and military occupation, reconstruction and redemption. Southerners, unlike other Americans, repeatedly felt the solid ground of continuity give way under their feet. An old order of slave society solidly supported by constitution, state, church and the authority of law and learning and cherished by a majority of the people collapsed, perished and disappeared. So did the short-lived experiment in national independence. So also the short-lived experiment in Radical Reconstruction. The succeeding order of Redeemers, the New South, lasted longer, but it too seems destined for the dump heap of history.

Perhaps it was because Cash wrote toward the end of the longest and most stable of these successive orders, the one that lasted from 1877 to the 1950's, that he acquired his conviction of stability and unchanging continuity. At any rate, he was fully persuaded that "the mind of the section . . . is continuous with the past," and that the South has "always marched away, as to this day it continues to do, from the present toward the past." Just as he guardedly conceded diversity in advancing the thesis of unity, so he admits the existence of change in maintaining the thesis of continuity, change from which even the elusive Southern "mind" did not "come off scot-free." But it was the sort of change the French have in mind in saying, *"Plus ça change, plus c'est la même chose."* Tidewater tobacco, up-country cotton, rampaging frontier, flush times in Alabama and Mississippi, slavery, secession, defeat, abolition, Reconstruction, New South, industrial revolution — *toujours la même chose!* Even the Yankee victory that "had smashed the Southern world" was "almost entirely illusory," since "it had left the essential Southern mind and will . . . entirely unshaken. Rather . . . it had operated enormously to fortify and confirm that mind and will." As for Reconstruction, again, "so far from having reconstructed the Southern mind in the large and in its essential character, it was still this Yankee's fate to have strengthened it almost beyond reckoning, and to have made it one of the most solidly established, one of the least reconstructible ever developed." [16]

The continuity upon which Cash is most insistent is the one he sees between the Old South and the New South. He early announces his intention of "disabusing our minds of two correlated legends — those of the Old and New South." He promises in Rankean terms to tell us "exactly what the Old South was really like." He concedes that there was a New South as well. "Nevertheless, the extent of the change and of the break between the Old South that was and the New South of our time has been vastly exaggerated." The common denominator, the homogenizing touchstone is his "basic Southerner" or "the man at the center." He is described as "an exceedingly simple fellow," most likely a hillbilly from the backcountry, but fundamentally he is a petit bourgeois always on the make, yet ever bemused by his vision of becoming, imitating, or at least serving the planter aristocrat. Cash's crude Irish parvenu is pictured as the prototype of the planter aristocrat. Cash is confused about these aristocrats, mainly I think because he is confused about the nature and history of aristocracy. He admires their "beautiful courtesy and dignity and gesturing grace," but deplores their "grotesque exaggeration" and their "pomposity" and suspects that the genuine article should have been genteel. He grudgingly acknowledges their existence, but denies the legitimacy of their pretenses — all save those of a few negligible Virginians. He seems to be saying

[16] Cash repeats the old-school picture of Reconstruction as "the brutal oppression of an honorably defeated and disarmed people."

that they were all bourgeois, that therefore the Old South was bourgeois too, and therefore essentially indistinguishable from the New South. New and Old alike were spellbound by the spurious myth of aristocracy. This and the paradoxical fact that those parvenu aristocrats actually took charge, were a real ruling class, and the continuity of their rule spelled the continuity of the New South with the Old.

The masses came out of the ordeal of Civil War with "a deep affection for these captains, a profound trust in them," a belief in the right "of the master class to ordain and command." And according to Cash, the old rulers continued to ordain and command right on through the collapse of the old order and the building of the new. He detects no change of guard at Redemption. So long as the industrialists and financiers who stepped into the shoes of the old rulers gave the Proto-Dorian password and adopted the old uniforms and gestures, he salutes them as the genuine article. In fact they were rather an improvement, for they represent "a striking extension of the so-called paternalism of the Old South: its passage in some fashion toward becoming a genuine paternalism." Cash enthusiastically embraces the thesis of Broadus Mitchell's "celebrated monograph" that the cotton-mill campaign was "a mighty folk movement," a philanthropic crusade of inspired paternalists.[17] The textile-mill captains were "such men as belonged more or less distinctively within the limits of the old ruling class, the progeny of the plantation." Indeed they were responsible for "the bringing over of the plantation into industry," the company town. Even "the worst labor sweaters" were "full of the ancient Southern love for the splendid gesture," fulfilling "an essential part of the Southern paternalistic tradition that it was an essential duty of the upper classes to look after the moral welfare of these people."

To the cotton mills the neopaternalists add the public schools for the common whites and thus "mightily reaffirm the Proto-

[17] Broadus Mitchell, *The Rise of the Cotton Mills in the South* (Baltimore, 1921).

Dorian bond." The common poverty acted as a leveler (back to the Unity thesis) and brought "a very great increase in the social solidarity of the South," a "marked mitigation of the haughtiness" of the old captains, now "less boldly patronizing," and "a suppression of class feeling that went beyond anything that even the Old South had known." The common white felt "the hand on the shoulder . . . the jests, the rallying, the stories . . . the confiding reminders of the Proto-Dorian bond of white men." That, according to Cash, was what did in the Populist revolt and the strikes of the lint-head mill hands as well. For from the heart of the masses came "a wide, diffuse gratefulness pouring out upon the cotton-mill baron; upon the old captains, upon all the captains and preachers of Progress; upon the ruling class as a whole for having embraced the doctrine and brought these things about."

Of course Cash professes not to be taken in by Progress like the red-necks and the lint-heads. He realizes that Progress and Success had their prices and he sets them down scrupulously in the debit column of his ledger. "Few people can ever have been confronted with a crueler dilemma" than the old planter turned supply merchant to his former huntin' and fishin' companion as sharecropper: "The old monotonous pellagra-and-rickets-breeding diet had at least been abundant? Strip it rigidly to fat-back, molasses, and cornbread, dole it out with an ever stingier hand . . . blind your eyes to peaked faces, seal up your ears to hungry whines. . . ." And that sunbonnet, straw-hat proletariat of the paternalistic mill villages? By the turn of the century they had become "a pretty distinct physical type . . . a dead white skin, a sunken chest, and stooping shoulders. . . . Chinless faces, microcephalic foreheads, rabbit teeth, goggling dead-fish eyes, rickety limbs, and stunted bodies. . . . The women were characteristically stringy-haired and limp of breast at twenty, and shrunken hags at thirty or forty." Something admittedly was happening to the captains, too, what with "men of generally coarser kind coming steadily to the front." And in "all the elaborate built-up pattern of leisure and hedonistic drift; all the slow, cool, gracious and graceful gesturing of movement," there was a sad falling off, a decay of the ideal. "And along with it, the vague largeness of outlook which was so essentially a part of the same aristocratic complex; the magnanimity . . ."

Admitting all that, "But when the whole of this debit score of Progress is taken into account, we still inevitably come back to the fact that its total effect was as I have said." *Plus ça change!* "Here in a word, was triumph for the Southern will . . . an enormous renewal of confidence in the general Southern way." In Grady's rhetoric, "Progress stood quite accurately for a sort of new charge at Gettysburg." To be sure, Southern Babbitts eventually appeared, but even they were "Tartarin, not Tartuffe . . . simpler, more naïve, less analytical than their compatriots in Babbittry at the North. . . . They go about making money . . . as boys go about stealing apples . . . in the high-hearted sense of being embarked upon capital sport." Yet, like the planter turned supply merchant or captain of industry, "they looked at you with level and proud gaze. The hallmark of their breed was identical with that of the masters of the Old South — a tremendous complacency." And Rotary, "sign-manual of the Yankee spirit"? Granting "an unfortunate decline in the dignity of the Southern manner," it was but "the grafting of Yankee backslapping upon the normal Southern geniality. . . . I am myself," Cash wrote, "indeed perpetually astonished to recall that Rotary was not invented in the South." And does one detect "strange notes — Yankee notes — in all this talk about the biggest factory, about bank clearings and car loadings and millions"? Strange? Not for Jack Cash. "But does anybody," he actually asked, "fail to hear once more the native accent of William L. Yancey and Barnwell Rhett, to glimpse again the waving plume of, say, Wade Hampton?"

How could he? How could any historian? He sometimes reminds one of those who scribble facetious graffiti on Roman ruins. He betrays a want of feeling for the seriousness of human strivings, for the tragic theme in history. Looking back from mid-twentieth century over the absurd skyscrapers and wrecked-car bone piles set in the red-clay hills, how could he seriously say that the South believed it "was succeeding in creating a world which, if it was not made altogether in the image of that old world, half-remembered and half-dreamed, shimmering there forever behind the fateful smoke of Sumter's guns, was yet sufficiently of a piece with it in essentials to be acceptable." A great slave society, by far the largest and richest of those that had existed in the New World since the sixteenth century, had grown up and miraculously flourished in the heart of a thoroughly bourgeois and partly puritanical republic. It had renounced its bourgeois origins and elaborated and painfully rationalized its institutional, legal, metaphysical, and religious defenses. It had produced leaders of skill, ingenuity, and strength who, unlike those of other slave societies, invested their honor and their lives, and not merely part of their capital, in that society. When the crisis came, they, unlike the others, chose to fight. It proved to be the death struggle of a society, which went down in ruins. And yet here is a historian who tells us that nothing essential changed. The ancient "mind," temperament, the aristocratic spirit, parvenu though he called it — call it what you will, *panache* perhaps — was perfectly preserved in a mythic amber. And so the present is continuous with the past, the ancient manifest in the new order, in Grady, Babbitt, Rotary, whatever, *c'est la même chose.*

I am afraid that Cash was taken in by the very myth he sought to explode — by the fancy-dress charade the New South put on in the cast-off finery of the old order, the cult of the Lost Cause, the Plantation Legend and the rest.[18] The new actors threw themselves into the old roles with spirit and conviction and put on what was for some a convincing performance. But Cash himself, even though he sometimes took the Snopeses for the Sartorises, plainly saw how they betrayed to the core and essence every tenet of the old code. "And yet," he can write,

[18] See above, pp. 44–45.

And yet — as regards the Southern mind, which is our theme, how essentially superficial and unrevolutionary remain the obvious changes; how certainly do these obvious changes take place within the ancient framework, and even sometimes contribute to the positive strengthening of the ancient pattern.

Look close at this scene as it stands in 1914. There is an atmosphere here, an air, shining from every word and deed. And the key to this atmosphere . . . is that familiar word without which it would be impossible to tell the story of the Old South, that familiar word "extravagant."

[Then, after a reference to the new skyscrapers in the clay hills:]

Softly; do you not hear behind that the gallop of Jeb Stuart's cavalrymen?

The answer is "No"! Not one ghostly echo of a gallop. And neither did Jack Cash. He only thought he did when he was bemused.

After some years in the profession, one has seen reputations of historians rise and fall. The books of Ulrich Phillips and later Frank Owsley began to collect dust on the shelves, and one thinks of Beard and Parrington. In America, historians, like politicians, are out as soon as they are down. There is no comfortable back bench, no House of Lords for them. It is a wasteful and rather brutal practice, unworthy of what Cash would agree are our best Southern traditions. I hope this will not happen to Cash. The man really had something to say, which is more than most, and he said it with passion and conviction and with style. Essentially what he had to say is something every historian eventually finds himself trying to say (always risking exaggeration) at some stage about every great historical subject. And that is that in spite of the revolution — any revolution — the English remain English, the French remain French, the Russians remain Russian, the Chinese remain Chinese — call them Elizabethans or Cromwellians, Royalists or Jacobeans, Czarists or Communists, Mandarins or Maoists. That was really what Cash, at his best, was saying about Southerners, and he said it better than anybody ever has — only he rather overdid the thing. But in that he was merely illustrating once more that ancient Southern trait that he summed up in the word "extravagant." And, for that matter, his critic, poured in the same mold, may have unintentionally added another illustration of the same trait. If so, Jack Cash would have been the first to understand and not the last to forgive. Peace to his troubled spirit.

Acknowledgments

While I have not imposed the whole manuscript of this book on any one friend, several have kindly read parts of it. All parts profit from the advice and criticism of one or more of them. I should like particularly to thank Edmund S. Morgan, C. R. Boxer, David Brion Davis, Willie Lee Rose, Philip D. Curtin, David M. Potter, Eugene D. Genovese, William S. McFeely, William G. Carleton, Sheldon Hackney, James M. McPherson, Norman Pollack, and Otto H. Olsen. All of them will recognize the influence their suggestions have had, though some may not agree with the uses made of them nor with the conclusions reached.

The theme of the book has been on my mind for about ten years, more especially since I moved from the South to New England in 1962, and has guided the writing of the parts previously published. All of these have been revised or enlarged, some of them extensively, some only in details. Three of the chapters are published here for the first time: "Protestant Slavery in a Catholic World," "Southern Slaves in the World of Thomas Malthus," and "The Strange Career of a Historical Controversy." Parts of the Preface appeared first in "The North and the South of It," *American Scholar,* XXXV (1966), 647–58. "The Southern Ethic in a Puritan World" appeared in *The William and Mary Quarterly,* XXV, Third Series (July, 1968), 343–70. "A Southern War Against Capitalism" served as an introduction to my edition of George Fitzhugh, *Cannibals All!, or Slaves Without Masters,* for the John Harvard Library (Cambridge, Massachusetts, 1960). "The Northern War Against Slavery" grew out of a review essay for the *New York Review of Books* (February 27, 1969), though it has been considerably enlarged and revised. "Seeds of Failure in Radical Race Policy" was originally published in *Proceedings of the American Philosophical Society,* CX (February, 1966), 1–9. "A Southern Brief for Racial Equality" was written as the introduction to my edition of Lewis H. Blair, *A Southern Prophecy: The Prosperity of the South Dependent Upon the Elevation of the Negro,* originally published in 1889 under the longer part of that title and reprinted by Little, Brown and Company in 1964. "The National Decision Against Equality" grew out of a piece originally published in *American Heritage,* XV (1964), 52–55, though it has been greatly revised. "The Elusive Mind of the South" appeared first in a slightly different version in the *New York Review of Books* (December 4, 1969). I wish to thank the publishers for permission to use these items in the present volume.

Index